U0332957

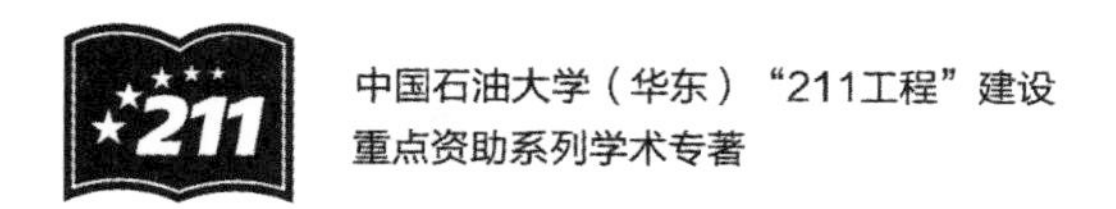

Interface: From Superplasticity to Atomic Phase Diagram

相界扩散溶解层

——从超塑性到原子相图

李世春　著

图书在版编目(CIP)数据

相界扩散溶解层:从超塑性到原子相图=
Interface: from superplasticity to atomic phase
diagram:英文/李世春著. 一东营:中国石油大学
出版社, 2015.3
ISBN 978-7-5636-3341-8

Ⅰ.①相… Ⅱ.①李… Ⅲ.①超塑性合金—研究—英
文 Ⅳ.①TG135

中国版本图书馆 CIP 数据核字(2015)第 054879 号

书　　名: Interface:From Superplasticity to Atomic Phase Diagram
相界扩散溶解层——从超塑性到原子相图
作　　者: 李世春

责任编辑: 袁超红(电话 0532—86981532)
封面设计: 悟本设计

出 版 者: 中国石油大学出版社(山东 东营　邮编 257061)
网　　址: http://www.uppbook.com.cn
电子信箱: shiyoujiaoyu@126.com
印 刷 者: 青岛国彩印刷有限公司
发 行 者: 中国石油大学出版社(电话 0532—86981531,86983437)
开　　本: 185 mm×260 mm　印张:12.25　字数:283 千字
版　　次: 2016 年 3 月第 1 版第 1 次印刷
定　　价: 78.00 元

Preface to the Series

"Project 211" approved by the State Council in 1995, is the largest and highest higher education development project established by the Chinese government. It is an important decision to the higher education development under the domestic and international situation at the turn of the century. Focusing on the core issues such as academic discipline establishment and teaching staff construction, "Project 211" is looking to bring major breakthrough to the overall development of higher education institutions, and to explore for a successful strategy in constructing higher caliber institutions. After 17 years effort, "Project 211" has made significant progress in higher education quality, scientific research, management and benefit improvement, and has laid a preliminary foundation for China's construction aim of a certain number of world-class universities.

Enlistment of China University of Petroleum (CUP) on the priority list of "Project 211" institution development in 1997, presented a historic opportunity for us to move forward towards the goal of building a high-level university. In the three construction periods of "Project 211" during the 9th, 10th and 11th Five-Year Plan, we always focus on the central theme of improving higher education, take satisfying the significant demands of petroleum and petrochemical industry as the mission, achieve key breakthrough of national oil and gas resources innovation platform as the goal, enhance the level of key disciplines, encourages academic leaders, cultivate international and innovative talent as the fundamental, adhere to the spirit of doing certain things and refraining from doing other things, try to achieve the overall development and improvement by existing advantages and characteristics. Through these efforts, we have enhanced the core competitiveness significantly, improved education level and comprehensive strength observably, and obtained a good foundation for the high-level research university with first-class petroleum discipline in world. After the construction of "Project 211", our petroleum and petrochemical characteristic is more distinctive, and the discipline advantages are more prominent. The constructions of innovation platforms of advantage disciplines progress well, and 5 national key disciplines and 2 national key disciplines (nurture) achieve the leading domestic and international advanced level. According to ESI's data updated in March 2012, engineering and chemical fields for the first time were evaluated into the ESI world rank. This reflects our obvious strength of main disciplines in petroleum and petrochemical. The construction of high-level teacher staff make substantial progress. A number of high-level talent teams such as Academicians, Changjiang Scholar Professor, winners of the National Science Fund for Distinguished Young Scholars, the Recruitment Program of Global Experts and experts of Baiqianwan Talents Project, provide a talent guarantee for the future development of the university. Scientific and technological innovation abilities are enhanced significantly, high-level projects and achievements emerge continually, scientific research funds exceed 400 million yuan every year. We have established initially the science and technology innovation system with distinctive petroleum characteristics, and become an important part of national innovation system. The ability to cultivates innovative talents is improved continuously, Excellent Engineer Education and Training Program and top-notch innovative talents cultivation zone are carried out. We actively explore the international talents training, deepen the reform of postgraduate training mechanism, and preliminarily establish our training mode and mechanism of graduate students that adapt to the cultivation of innovative talents. We improve the public service system construction continuously, and have build the advanced and efficient public service system. The software and hardware conditions of university are improved significantly, which is a strong guarantee for the teaching, scientific research and management level.

The track of 17 years "Project 211" developments marks the important theme and sign of our university's development. The experiences gained from "Project 211" construction become a valuable asset to us. Firstly, we must insist on the spirit of doing certain things and refraining from doing other things, achieve breakthroughs in a few disciplines by strengthening the characteristics and outstanding advantages, and then strive to achieve the international first-class petroleum development goals. Secondly, we need to adhere to the rolling development and overall improvement, and drive overall development by developing the key, further expand our advantages and pursue coordinated development, in order to constantly improve the overall competitiveness. Thirdly, we must insist on perfecting the mechanism and building a platform, through the improvement of operation mechanism of "united, open, sharing, competition, flow" and the platform construction mechanism based on projects, strengthen overall planning and resource integration, centralize power of talents, optimize the construction process and working system, in order to ensure all the work efficient and orderly conduct. Fourthly, we must adhere to enhancing the talent pool and concerted efforts, train and gather a large number of outstanding talents to form a dedicated and innovative team through the promotion of "Project 211" construction and the development in all aspects, form a powerful force by harmony, unity and cooperation of colleges, disciplines and related departments, and then ensure the smooth implementation of the construction tasks. These experiences are formed in the long-term practice of the "Project 211" construction, and we must be better to inherit and carry forward in order to further promote the construction and development of high-level research universities.

For the sake of concluding the successful experiences gained from "Project 211" construction, fully displaying the rich fruits of "Project 211" construction, we set up the special fund since 2008 to publish a series of academic monographs which are related to "Project 211" construction. Those academic monographs are based on the scientific research achievements of CUP's outstanding scholars, introduce and demonstrate the achievements and experiences of the discipline construction, technological innovation and personnel training and so on. We believe this series can further inherit the advanced achievements of scientific research and academic thoughts, and show the great achievements and development ideas of "Project 211" construction from different aspects, so as to play a unique contribution and role to expand our influence in society, improve academic reputation, and promote the future "Project 211" construction.

Finally, thanks to the majority of scholars for their hard work and great efforts to CUP's "Project 211" construction. They diligently summarize the monograph research achievements, and leave the precious innovations and academic spirit for academic career, teachers and students.

President, China University of Petroleum(East China)

September 2012

16 years ago, the Division I of Materials Science of National Natural Science Foundation of China gave me a seed, and by now the seed has grown into seedling. When will it blossom, I do not know; when will it bear fruit, I also do not know. The story below is the process of growing a seed into a seedling.

When I obtained my master's degree in 1988 and came back to China University of Petroleum (at Dongying), I started applying for funding from National Natural Science Foundation of China (NSFC). I knew the probability of success at that time was only about 15% and the competition is nation-wide. But if I do not apply the probability of getting funding is always zero, and if I applied the probability of getting funding is greater or equal to zero. The thing to emphasize is that, it means success only when the probability is equal to one. It is like this when applying for funding and also for realizing the final goal of a research project.

From close to 20 years experience of trial and error, I learned the three requirements of doing basic research, which are: scientific thought, scientific spirit, and scientific method. Now with examples from my own experience, I will discuss three reflections modeling from "three states of reading" by Wang Guowei.

1. Deciding the research subject

Last night the west wind withered the green trees.
Alone I climbed the high pavilion,
Gazing at the distant road vanishing into the horizon.

—Song Prose, by Yan Shu (991—1055)

Translated by Yang Xianyi and Dai Naidie.

After obtaining my master's degree in 1988 (the title of my master's thesis is *The effects of rare earth metals on the superplasticity of Al-Zn-Mg alloys*), I sensed that superplasticity, NSFC, and Tsinghua University are somehow related. Firstly, I will use superplasticity to apply for a project from NSFC; Secondly, I will use superplasticity to apply for a PhD degree at Tsinghua University.

It took me 5 years to get the first NSFC project, until 1993 the fifth proposal was approved; it took me 12 years to obtain a PhD degree from Tsinghua University, until

2000 I finished my oral defense.

In December 1991, I knew a professor, Mr. Chen Nanping, at Tsinghua University, and hoped that he could aid my research in superplasticity. I asked him to write me a specialist recommendation letter for my NSFC proposal. Professor Chen wrote the letter, but the proposal still did not get approved. Before this, my application for NSFC funding had been rejected three times, so the probability of success for these four applications are all zero.

In 1992, I published a book with China Press of Broadcast & Television. The title of the book is *Rubik's Cube and Its Application*. After writing about Rubik's cube, my proposal for NSFC project included some ideas of a kick in one's gallop including the problem of Rubik's cube.

In July 1993, before all the departments of NSFC organized the project evaluation meetings, I had mailed the pre-prepared three copies of *Rubik's Cube and Its Application* to Division I of Materials Science of NSFC, one for each administrator, and people say that the Division head brought my book to the project evaluation meeting.

In October 1993, as a lecturer at China University of Petroleum, I obtained my first NSFC funded project, *Dissipative Structural Model of Superplasticity and Its Research in Metal Physics*, with Rubik's cube and the heavily modified proposal.

At the same time, I was introduced to Cheng Kaijia, winner of the Two Bombs and One Satellite Award and member of the Chinese Academy of Science, to study the TFDC model. I have being focusing on TFDC electron density.

I researched superplasticity for 12 years, getting NSFC funding twice, and the main result is getting a "diffusion-dissolution zone model", at the same time achieved the transition from superplasticity to TFDC model. After that, I had the idea of atomic phase diagram, which is a theoretical model for atomic assembly technology in the future. This theoretical model can be called the TFDC atomic assembly theory and the initial results were published in the *Progress in Natural Science* in 2004. The final goal of my research will be establishing the TFDC atomic assembly theory, which is centered on the concept of TFDC electron density and its applications.

2. Maintaining the research subject

My clothes hang loose on my emaciated body,
But regrets I have none, it is because of her.

—Song Prose, by Liu Yong(987—1053?)

Translated by Yang Xianyi and Dai Naidie.

With my master's degree, I worked as a lecturer of China University of Petroleum located in the city of Dongying at the delta of the Yellow River. I knew from the very start that the lack of instrumentation and funding would cause some physical hardship with the byproduct of some wisdom. This is the equivalence principle of nature.

During the time as a graduate student, I learned a set of skills on my own, which is the use of hand saws, files, and bench clamps to prepare samples for stretching tests. Upon graduation and return to China University of Petroleum in 1988, I often visited a storage place for old and broken equipments and sometimes was able to get usable equipments. 4 years later, I made a superplasticity-testing machine using the gear system from a bicycle and 4 box-like ovens. The first try on a Zn-5Al alloy sample gave an elongation of 5000%.

From 1993 to 2000, I spend 8 years to explain the experimental result of 5000% and finally obtaining a "diffusion-dissolution zone model".

When studying the TFDC model of Professor Cheng Kaijia, I never gave up my chosen topic of superplasticity, because my second goal, which was obtaining PhD degree from Tsinghua University, has not yet been realized. Moreover, I did not directly participate in the research group of Professor Cheng, but instead wanted to make the transition from superplasticity to TFDC model in my own way. More than 10 years has past since I realized my ideas and achieved my transitional goals.

During these years, I looked up more than 3000 articles concerning TFD electron density involving many journals since the first day of publishing. Some examples include, *Physical Review Series* (1893 to now), *Journal of Physical Chemistry* (1896 to now), *Journal of Applied Physics* (1931 to now), *Journal of Chemical Physics* (1933 to now), *Advances in Physics* (1952 to now), and *Journal of Physics* (1968 to now). (This work was finished at UBC library Vancouver.)

I now have a rather complete understanding of the development of the concept of electron density, and also realized that the major originality of Professor Cheng Kaijia's TFDC model is the concept of TFDC electron density.

To sum it up: scientific thought is the soul, instrumentation and funding are at the most some flesh, but the scientific spirit is the backbone.

3. Deepening the research subject

A thousand times I search for her in the crowd.
And, suddenly turning my head,
Discover her where the lantern lights are dim.

—Song Prose, by Xin Qiji(1140—1207)

Translated by Yang Xianyi and Dai Naidie.

From 1988 when I graduated with the master's degree, performing scientific research independently for me is like crossing the river by feeling the rocks in the dark, without any light towers, without guides, and without any sound from the other side of the shore. During these times, I still hoped of meeting some real great masters in the same field, in order to communicate ideas and get some kind of help.

In 1993, when I first met Professor Cheng Kaijia, I only know that Professor

Cheng wrote a textbook in solid state physics. When I learned that TFDC model is the area of Professor Cheng's daily research for over 50 years, intuitions told me this is an opportunity, because the Division I of Material Science of NSFC will organize the personnel of the 6 general programs to participate in a 3-year academic activity.

In 1998, the *Guide to NSFC Programs* clearly showed Cheng's theory as a key program called *The Effects of Interface Electron Distribution in Multi-layer Material on the Properties of the Material*. I wrote an *Introduction to TFDC Model* with 30 pages as appendix for the proposal, which discussed how to develop the TFDC model. Therefore, I obtained the third NSFC project; the title also is *The Effects of Interface Electron Distribution in Multi-layer Material on the Properties of the Material*.

On one hand, I was writing my PhD dissertation based on research concerning superplasticity; on the other hand, I was trying to unlock the central idea of *The Effects of Interface Electron Distribution in Multi-layer Material on the Properties of the Material*. It is very important that the TFDC electron density can describe the contact interface of the heterogeneous atoms. However, the multi-layer materials contain different kinds of atoms in different layers.

As a side note, the Zn-Al eutectic alloy used for my superplasticity research has layered microscopic structure. The zone in which Al phase and Zn phase are in contact is the interface layer, and thickness of this interface layer can vary under the effects of diffusion and dissolution.

After much thought, the word "layer" is right here in the title. The answer to the problem I thought for many years is just the word "layer". The problem troubled me since 1993, which is the transition from superplasticity to TFDC model, the key to the transition is a bridge, and at this time it is very clear to me that the bridge is based on "layer".

In 2000, I combined the experimental results during the years and the word "layer" in my thinking, and wrote the results into my PhD dissertation, which finally formed the central idea of "diffusion-dissolution zone", and was emphasized throughout the dissertation. The dissertation was perused by Professor Chen Nanping of Tsinghua University, which later became the final version. I used 4 more years to extend the meaning of "diffusion dissolution zone", and expand the PhD dissertation into a book, *Diffusion Dissolution Zone in Interface*, and the Chinese version was published in 2006 with funding from the publishing foundation of NSFC.

After meeting Professor Cheng Kaijia, I became very interested in his TFDC electron density. After explaining several experimental results with TFDC electron density, I began to focus my attention on the meaning and originality of TFDC electron density through searching the original articles. The major question is what is the significance of TFDC electron density?

I realized that an atomic assembly theory could be formulated from TFDC electron density. People first made bricks from sand and soil, then use bricks to build houses. By the same token, first assemble the atoms into "atomic bricks" and predict some parameters of the "atomic bricks", then use them to build materials or nano device whose technical properties are predicted by some models.

In 1993, the Division Ⅰ of Material Science of NSFC introduced me to Professor Cheng Kaijia, winner of the Two Bombs and One Satellite Award and member of the Chinese Academy of Science through cooperated management method of general program, which gave me a chance to discuss the TFDC model with Professor Cheng. If this kind of opportunity is a seed, then the TFDC atomic assembly theory I am currently working on is a tender seedling.

When will the seedling blossom and bear fruit? It could be many years later!

(The Chinese version of this preface has been published in Science Foundation in China, Vol.21, No.2, 2007.)

Li Shichun

April 20, 2015

Chapter 1

Diffusion Dissolution Zone in Interface

When two different grains A and B are in contact and bonds together, an interface is formed between particle A and particle B, as shown in Figure 1.1. The grains form a system containing an interface and the binary particle system is used to describe such a system. Two solid particles in touch bond at the interface through diffusion and dissolution, thus diffusion and dissolution forms a phase interface at the boundary between the two particles.

(a) Particle A and particle B (b) A/B interface between A and B

Figure 1.1 Particles A and B before and after contact.

If the binary particle system with interface is subjected to high temperature (e.g. $0.5T_m$ where T_m is the melting point of the particle with lower melting point of the two) and kept at the temperature for a period of time, then diffusion and dissolution will occur at the interface of the binary particle system. The result is a distinct region forming in between the two particles, as shown in Figure 1.2, and its microstructure including the distribution of elements is different from both particle A and B. The spatial characteristic of the region is that it is divided into zones, thus the region is called the diffusion dissolution zone.

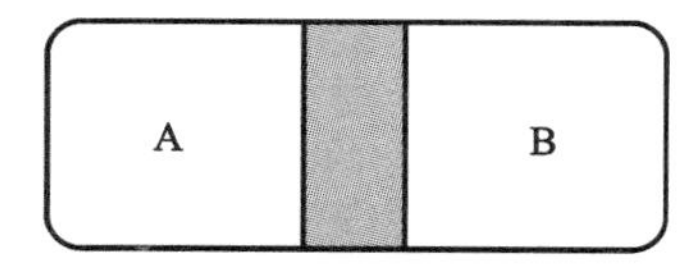

Figure 1.2 The diffusion dissolution zone between particles A and B.

Diffusion dissolution zone is a very common phenomenon in material science and its basic cause is the attraction and repulsion interactions between heterogeneous atoms. The binary alloy phase diagram in material science is just a representation of the diffusion dissolution zone between two components.

Phase diagrams are the most basic information in material science, which show not only the effect of temperature on the interaction between elements but also the formation of interface between elements. There are a lot of alloy phases in materials science and chemical elements are the basic units in alloy phase. The Periodic Table is the guide to the material world and it is also the guide to phase diagrams of materials. For example, if the elements in the Periodic Table are paired up, the different kinds of binary phase diagrams will be got. In fact, much of the information about phase diagrams and interface from the Periodic Table can been predicted.

1.1 The definition of diffusion dissolution zone

In 1978, Chinese physicist Yu Ruihuang formulated the "Empirical Electron Theory of Solid and Molecules (EET)"[1-2]. In this theory, first construct two atomic states or so-called h state and t state, then get hybridized states of atoms according to the corresponding formula and distribute the electrons among the bonds with the aid of the information of crystal space group, and finally calculate bond length using revised Pauling formula and obtain so-called theoretical bond length.

On the other hand, calculate some distances between atoms according to the lattice constant and obtain so-called experimental bond length. At last make a comparison between the theoretical bond length and the experimental bond length, if the difference is less than 0.05 angstrom, believe that constructed atomic states, i. e. electron structures are reasonable, otherwise, carry out the above calculation again until theoretical bond length and experimental bond length are in good agreement.

Yu Ruihuang defined the h state and t state of atoms in solids as the basic states and the other states are all obtained from combinations of the h and t states. In Yu's theory, h and t states are referred to as the head and tail states respectively. All states between the h and t states can be represented as combinations of the h and t states thus satisfying the superposition principle of quantum state.

Yu's symbols have been used for describing the interaction between atoms[3]. For two-atom system, h represents the uncontact state and t represents the contact state. For two-particle system, h represents the start state before a process and t represents the end state. The transition process from h to t state or the changes of a system from h to t state is the focus in this chapter.

1.1.1 h state and t state

In order to describe the contact between atoms, a concept of lattice atom has been introduced[3]. A lattice atom is the atom in h state, which is isolated from its crystal, but has the same electron structure as that in the crystal. Therefore, lattice atoms not in contact to be in h state and lattice atoms in contact to be t state. The process of atoms not in contact come into contact is a typical case of microstructure process in atomic scale that to date cannot be observed directly. The use of these definitions can describe not only atomic process but also macroscopic process and their differences are illustrated in Figures 1.3 and 1.4.

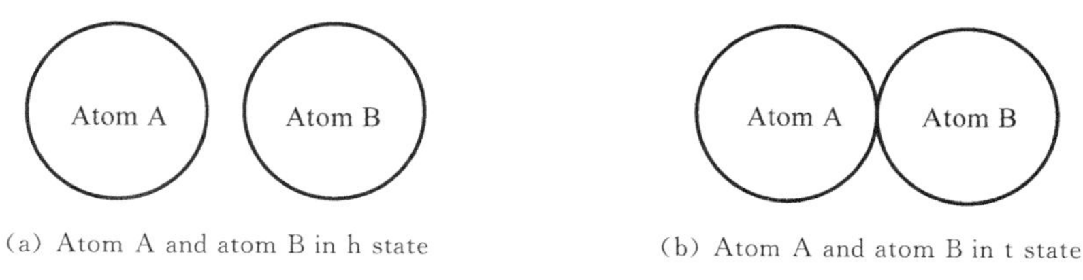

(a) Atom A and atom B in h state (b) Atom A and atom B in t state

Figure 1.3 The h state and t state of the atoms.

For a binary particle system, there are two particles before contact, after contact an interface is formed and now the two particles are in h state. When some kind of heat treatment is used to facilitate the development of the interface into a diffusion dissolution zone, now the two particles are in t state as illustrated in Figure 1.4.

(a) Particle A and particle B in h state (b) Particle A and particle B in t state

Figure 1.4 The particles in h state and t state.

The formation of the interface between components A and B under room temperature achieves the h state; the system is then subjected to high temperature (in solid state) and kept for a long enough period of time to let the interface develop into a diffusion dissolution zone, called t state. Through diffusion and dissolution, the system changes from h state to t state, the process is the diffusion and dissolution process with respect to the formation of the diffusion dissolution zone between components A and B.

1.1.2 Alloy phase diagram and diffusion dissolution zone

Points in alloy phase diagrams can describe the thermodynamic state of the system; a vertical line joining two points represents a change in temperature and can be called a vertical process, as shown in Figure 1.5. For example, the crystallization or melting process of a particular alloy is a vertical process, general textbooks usually contains a detail discussion of this topic.

In phase diagrams, the simplest horizontal process involves only one zone in the diffusion and dissolution process formed from h state to t state of a system composed of components A and B.

For example, the Ni-Ti binary phase diagram is shown in Figure 1.6. At between

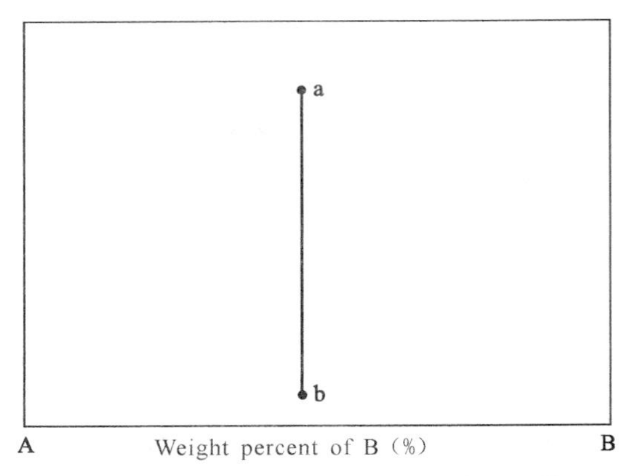

Figure 1.5 The vertical process in a phase diagram.

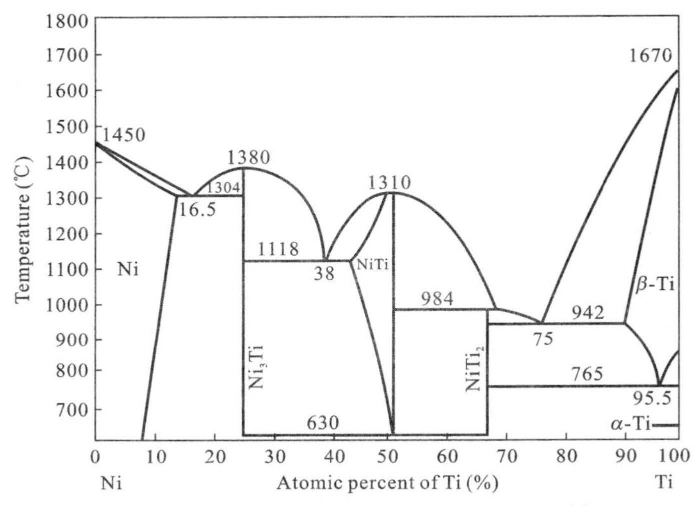

Figure 1.6 The Ni-Ti binary phase diagram[4].

630 ℃ and 765 ℃, there are five phases in the region from Ni to Ti: Ni, Ni_3Ti, NiTi, $NiTi_2$ and Ti. Ni and Ti are considered as two particles, and they come into contact to achieve the h state as shown in Figure 1.7. When the Ni/Ti in h state is kept at between 630 ℃ and 765 ℃ for 50 hours or more, a diffusion dissolution zone forms at the interface between Ni and Ti. The structure of the diffusion dissolution zone is: Ni phase, interface Ni/Ni_3Ti, Ni_3Ti phase, interface $Ni_3Ti/NiTi$, NiTi phase, interface $NiTi/NiTi_2$, $NiTi_2$ phase, interface $NiTi_2/Ti$ and Ti phase, as shown in Figure 1.8.

From the comparison between Figures 1.7 and 1.8, heat treatment produces a zone containing intermetallic compounds Ni_3Ti, NiTi and $NiTi_2$ at the interface Ni/Ti. These compounds are distributed in layers and are formed mainly by diffusion and dissolution, thus this zone is called diffusion dissolution zone.

The definition of diffusion dissolution zone was first presented in a PhD dissertation discussing superplastic mechanisms in Zn-Al eutectic alloy in the year 2000[5]. The content of the *Phase boundary sliding model controlled by diffusion solution zone in*

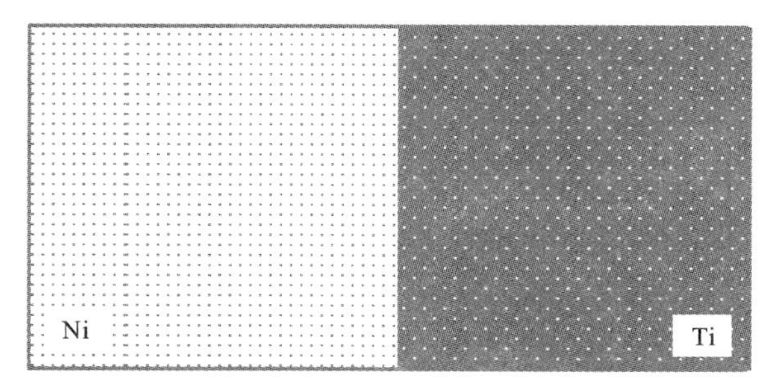

Figure 1.7 h state composed of Ni and Ti.

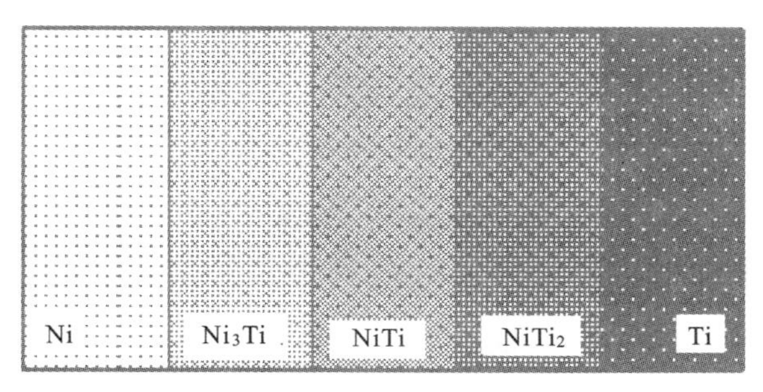

Figure 1.8 t state composed of Ni and Ti.

superplastic deformation published in *Chinese Science Bulletin* in March 2002 are entirely from Chapter 5 of this PhD dissertation, and the "author" of the article did not understand the meaning of diffusion dissolution zone. *Chinese Science Bulletin* (P1355, Vol. 49, No.14) explained the situation in July 2004[6].

In 2003, a research group from University of California at Davis, published an article in *Acta Materialia* with the title *Enhanced growth of intermetallic phases in the Ni-Ti system by current effects*[7]. In the article, there is a back-scattered electron micrograph taken with SEM as shown in Figure 1.9. The Ni_3Ti, NiTi and $NiTi_2$ zone in the figure, which is formed through diffusion into each other following initial contact, is a typical case of diffusion dissolution zone.

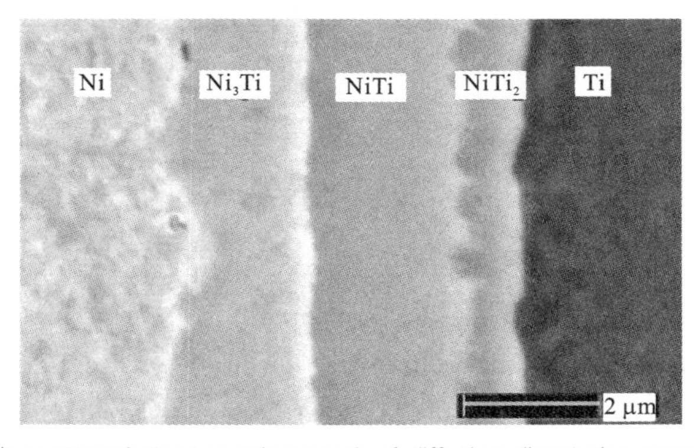

Figure 1.9 Back-scattered electron micrograph of diffusion dissolution zone of Ni and Ti[7].

1.1.3 Structure of diffusion dissolution zone

Two horizontally distributed points in a phase diagram can represent two different systems as shown in Figure 1.10. a′ and b′ in the figure are two points on the same horizontal line and also correspond to composition a and b on the horizontal axis respectively. If two alloys with compositions a and b respectively come in contact and go through diffusion treatment, then a diffusion dissolution zone will form at the interface where the two alloys come into contact. The phases included in the diffusion dissolution zone and their distributions are represented by the shaded area in the phase diagram. Thus, the diffusion dissolution zone between components A and B corresponds to the arrangement of their binary phase diagram.

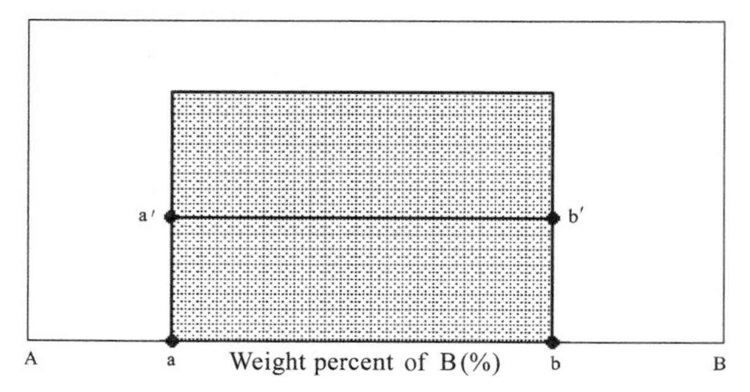

Figure 1.10 The horizontally distribution in a phase diagram.

For example, Ni_3Ti and $NiTi_2$ are used to form the h state as shown in Figure 1.11. After the Ni_3Ti and $NiTi_2$ in h state are subjected to diffusion anneal at between 630 ℃ and 765 ℃, a NiTi diffusion dissolution zone is formed at the $Ni_3Ti/NiTi_2$ interface, as shown in Figure 1.12.

Figure 1.11 h state composed of Ni_3Ti and $NiTi_2$.

In fact, Figure 1.12 and the central region of Figure 1.9 are exactly the same. This demonstrates that for any A-B phase diagram, if the compositions of α and β phases are a and b respectively, as shown in Figure 1.13, h state formed by α and β will have a diffusion dissolution zone after diffusion treatment, as shown in Figure 1.14. The structure of the diffusion dissolution zone ab and its phase arrangement will be the same as the phase arrangement of the region between α and β in A-B phase diagram.

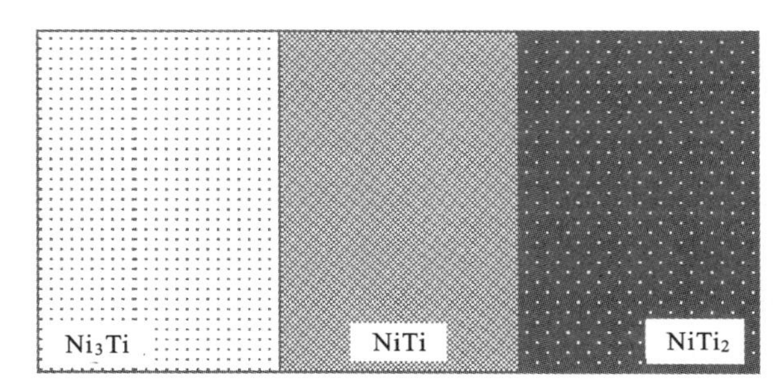

Figure 1.12 t state composed of Ni_3Ti and $NiTi_2$.

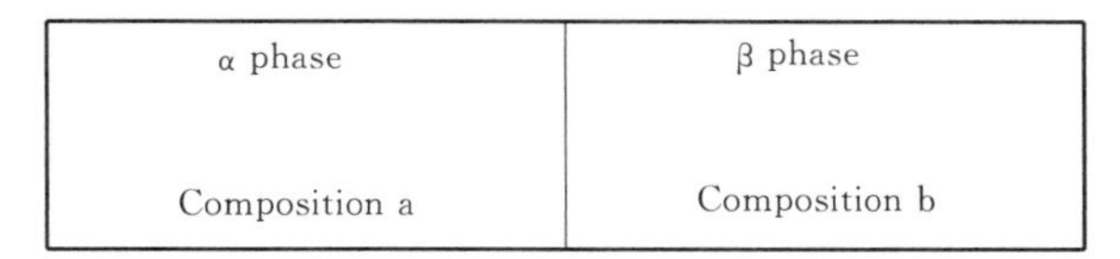

Figure 1.13 h state composed of α and β.

α phase		β phase
Composition a	Zone ab in phase diagram	Composition b

Figure 1.14 t state composed of α and β.

1.1.4 Catholicity of diffusion dissolution zone

Diffusion dissolution zone is a general physical phenomenon. Any two different particles, for example particles A and B, can form a diffusion dissolution zone after diffusion treatment. Before heat treatment the two particles are in h state and the boundary between the two particles is an interface; after heat treatment the two particles are in t state and the interface has been developed into a diffusion dissolution zone. The driving force for the formation of diffusion dissolution zone is chemical potential, whose fundamental cause is the interaction between valence shell electrons of two atoms.

The Periodic Table describes the electronic structure of free atoms. When the atoms interact with each other and form solid, the valence shell electrons of one atom will feel the effect of neighbouring atoms and chemical bond theory describes these kinds of effects. When different kinds of atoms interact with each other, the effects to the valence shell of different atoms will be different.

There are large quantities of heterogeneous atoms on both sides of interface, thus diffusion dissolution zone is formed by heterogeneous atoms. If a model that characterizes the interaction between two heterogeneous atoms can be formulated, then the model can be applied to interface and diffusion dissolution zone. The next section will discuss this topic.

1.2 Interface between atoms and atomic phase diagram

Classic phase diagram describes the equilibrium characteristic of alloy phase in equilibrium state and is commonly presented as temperature versus composition diagram. In two-phase zone, the relative abundance can be calculated according to the lever law. Atomic phase diagram tries to describe the equilibrium character of heterogeneous atoms in equilibrium state and is presented as electron density versus atomic radius diagram. The equilibrium electron density of two atoms in equilibrium state can also be calculated according to a lever law. The electron density versus atomic radius diagram is based on the ideas of TFDC model, thus atomic phase diagram can also be referred to as TFDC phase diagram.

The difference between atomic phase diagram and classic phase diagram is that atomic phase diagram only involves two states of an atom, which are the uncontact state and contact state. The so-called lattice atoms before contact are described by the classic TFD model and the state after contact is described by atomic phase diagram[3].

1.2.1 State of atom

The different kinds of data from the Periodic Table describe atoms in different states. For example, electron configurations apply to free atoms and atomic radii apply to atoms in crystals. In order to use the atomic phase diagram data, the two basic states of an atom defined in Section 1.1.1 are going to be described in detail here. When atoms are not in contact, they are in h state and when they are in contact, they are said to be in t state. Here the atomic state symbols used by Dr. Yu are borrowed to have completely different meanings. The TFDC model and Yu's theory were bridged by a concept of atomic interaction volume[8] on the level of atoms.

For simplicity's sake, assume component A and component B both can form its own kind of crystal, namely crystal A and crystal B respectively as shown in Figure 1.15, and this kind of assumption does not affect general applications.

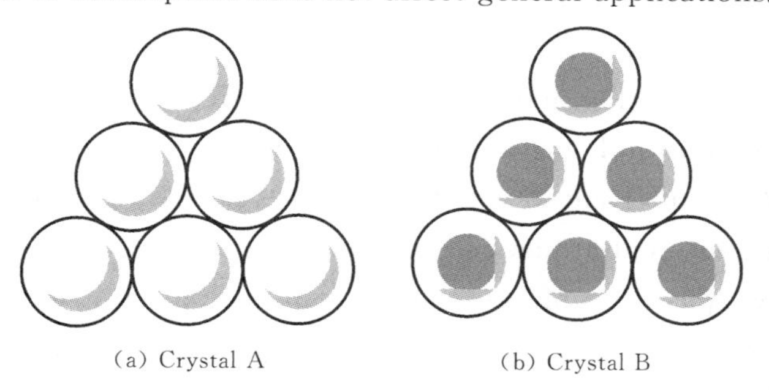

(a) Crystal A (b) Crystal B

Figure 1.15 Crystal state of atoms A and B.

Take one atom of kind A and kind B from the lattice environment in Figure 1.15, and put them under h state as shown in Figure 1.16. The isolated atom in h state is completely different from a free atom. Although isolated from the lattice environment formed by homogenous atoms, the electronic configurations of atoms in h state remains unchanged and thus can be called a lattice atom. Such an atom in h state does not exist in reality but the later theories need a definition like this.

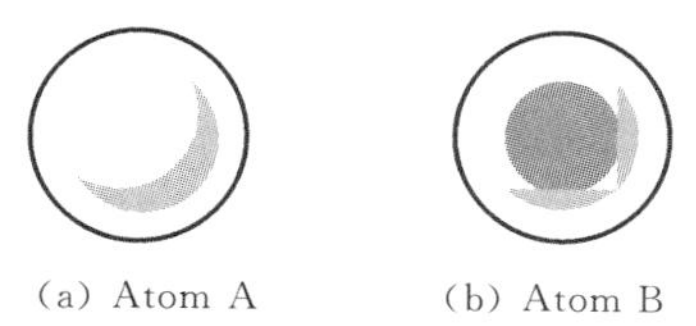

(a) Atom A (b) Atom B

Figure 1.16 h state of atoms A and B.

When an atom of type A (atom A) and an atom of type B (atom B) in h states come into contact, the electron densities of atom A and atom B need to be rearranged in order to make the electron densities equal at the interface between the two atoms. In order to satisfy the requirement that the electron density is continuous at the interface region, the two atoms would adjust their electronic structure and reside in a joint state. This kind of state can be denoted as t state, as shown in Figure 1.17.

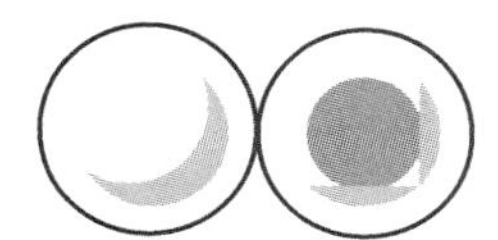

Figure 1.17 t state of atoms A and B.

1.2.2 Data of electron density

Thomas[9] and Fermi[10] formulated a model describing the electron distribution of atoms called TF model. Its mathematical expression is as below

$$n(x)=\frac{z}{4\pi\mu^3}\left(\frac{\Phi}{x}\right)^{\frac{3}{2}} \tag{1-1}$$

where $\mu = a_0\left(\frac{9\pi^2}{128z}\right)^{1/3}$, a_0 is Bohr's radius, z is the atomic number, Φ is the dimensionless TF function, x is the dimensionless atomic radius, $r=\mu x$ is the actual atomic radius, $n(x)$ is the electron density.

In 1930, Dirac[11] introduced electron exchange act into the TF model and formulated the TFD model. TFD model gives the relationship between electron density and atomic radius as the following

$$n(x)=\frac{z}{4\pi\mu^3}\left[\varepsilon+\left(\frac{\Psi}{x}\right)^{\frac{1}{2}}\right]^3 \tag{1-2}$$

where ε is the electron exchange term introduced by Dirac, $\varepsilon=\left(\frac{3}{32\pi^2}\right)^{1/3}z^{-2/3}$, Ψ is called TFD function, which satisfies the famous TFD equation.

$$\frac{d^2\Psi}{dx^2}=x\left[\varepsilon+\left(\frac{\Psi}{x}\right)^{\frac{1}{2}}\right]^3 \qquad (1\text{-}3)$$

In fact, the equation is derived from the Poisson equation, thus the assumption that electron clouds are distributed with spherical symmetry is made. TFD model is applied successfully to the problem of solving the atomic scattering factor in X-ray diffraction. When given the atomic radius, one can use Equations (1-2) and (1-3) to solve the electron density. Dr. Cheng Kaijia worked on this kind of problems as early as 1993, and formally published the electron density data of 38 elements in the Periodic Table in 1996[12].

One thing to note is that the electron density solved from TFD model is in fact the n_{WS} in literature. It is documented in literature[13] in 1971 that people tried to use TFDW to solve electron density at the Wigner-Seitz radius, or n_{WS}, but due to the incorrect selection of boundary conditions they only got the graphical relationship of the electron density increasing with the radius without specific data. Before Dr. Cheng Kaijia published his results, there are no literatures documenting the electron density data at the Wigner-Seitz radius calculated by the classic TFD model. This is a new contribution to the classic TFD model by Dr. Cheng. The electron density data Dr. Cheng calculated using TFD model could be regarded as a new parameter for elements.

In 1973, Miedema[14] formulated an empirical formula based on the valence of alkali metals and atomic volume data.

$$n_{WS}=0.82\times10^{-4}(B/V_m)^{1/2} \qquad (1\text{-}4)$$

where B is the bulk module with unit kg · cm^{-2}, V_m is the atomic volume with unit of cm^3, n_{WS} is the electron density at the Wigner-Seitz radius with unit electron number of (a.u.)^3.

Miedema has given the electron density data of 24 elements based on Equation (1-4). In 2001, Ye Hengqiang's group of Institute of Metal Research, Chinese Academy of Sciences calculated the electron density data of 26 metals using the LAPW method based on local density function[15], and compared the data to that of Miedema's formula, as shown in Table 1.1. One thing to point out is that the data from references [15] and [14] is almost the same, but this does not indicated that the theory proposed in article [15] agrees with experimental results since the data from article [14] are estimated using an empirical formula. It is possible that the method used in article [15] includes some parameters that are decided with reference to the data in article [14] which causes the results to be almost same.

Table 1.1 TFD electron density, Miedema electron density and Ye Hengqiang electron density

Z	Name	r	n_{TFD}	n_{WS}[14]	n[15]
3	Li	1.7316	0.24	0.53	0.52
11	Na	2.11	0.18	0.30	0.29

Continued

Z	Name	r	n_{TFD}	n_{WS}[14]	n[15]
13	Al	1.5825	1.07	1.51	1.73
19	K	2.6187	0.07	0.15	0.16
20	Ca	2.28	0.15	0.41	0.50
23	V	1.4901	2.04	2.52	2.49
24	Cr	1.4202	2.64	2.91	2.94
26	Fe	1.4119	2.85	2.74	2.62
28	Ni	1.378	3.34	3.01	2.66
29	Cu	1.4119	3.03	2.43	2.20
37	Rb	2.8088	0.07	0.13	0.13
38	Sr	2.3728	0.17	0.34	0.39
41	Nb	1.6237	1.79	2.20	2.36
42	Mo	1.5504	2.31	3.05	2.89
45	Rh	1.4873	2.95	3.24	2.88
46	Pd	1.5224	2.65	2.53	2.30
47	Ag	1.5982	2.09	1.73	1.65
55	Cs	3.0275	0.07	0.10	0.10
56	Ba	2.4912	0.14	0.27	0.38
73	Ta	1.6288	2.33	2.44	2.51
74	W	1.5573	2.97	3.29	3.18
77	Ir	1.5015	3.66	3.66	3.34
78	Pt	1.5336	3.29	3.15	2.78
79	Au	1.5932	2.71	2.33	2.09
82	Pb	1.9359	0.94	0.89	1.00

Note: Atom radius in 10^{-10} m, the unit of electron density in $10^{23}/m^3$.

One thing to emphasize is that the boundary condition of Equation (1-3) is

$$\Psi(0)=1 \tag{1-5}$$

$$x_0 \frac{d\Psi(x_0)}{dx_0}=\Psi(x_0) \tag{1-6}$$

For a given atomic radius r_0, solving Equation (1-3) first followed by Equation (1-2) gives a curve describing the electron density distribution or n'-r'; when the atomic radius changes from r_0 to $r_0+\Delta r$, the electron density distribution curve is completely different, i.e. n''-r'', which is not a simple extension of the n'-r' curve, as shown in Figure 1.18.

In conclusion, TFD model is able to describe the state of atoms; the state parameters are atomic radii and electron densities. If the atomic radii (atomic volumes) are the atomic radii (atomic volumes) in a crystal, then it is the h state of atoms defined above.

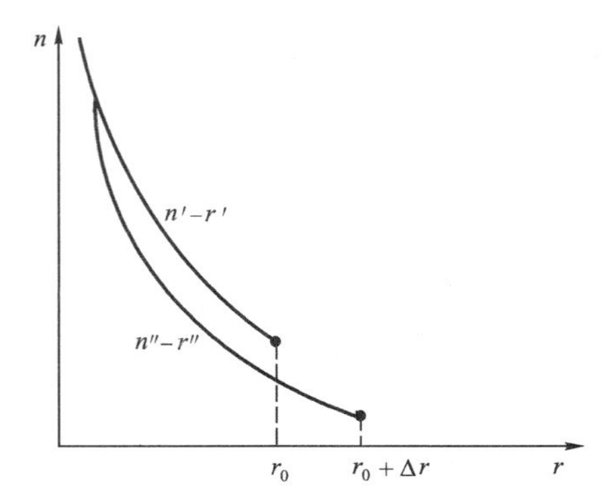

Figure 1.18 Electron density vs. atomic radius.

1.2.3 Atom contact between heterogeneous atoms

When two heterogeneous atoms go from the uncontact h state described using the classic TFD model to the t state described using atomic phase diagram, the process underwent includes two physics mechanism: the first is the transfer of valence electron (in fact, the transfer of valence means a abruptly change in the electron density in the valence shell of atoms) and the second is the change of electron density in the valence shell of atoms.

When two components form solid solution or a compound, the surface electron density of atom of each component will change in order for the electron density to be continuous at the two sides of the interface. The atom with higher electron density will decrease its electron density while the atom with lower electron density will increase its electron density. For example, Figures 1.19 and 1.20 are diagrams illustrating the continuity requirement of electron density at the interface between atoms of TFDC model. There are four curves of electron density versus atomic radius in the each of the diagrams, i. e. the *n*-*r* curves, which are the electron density distribution of atoms calculated using the classical TFD model with given atomic radii.

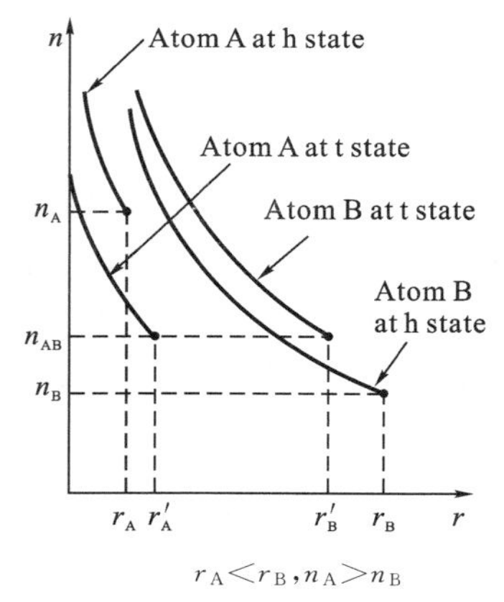

Figure 1.19 Electron density continuity between atoms.

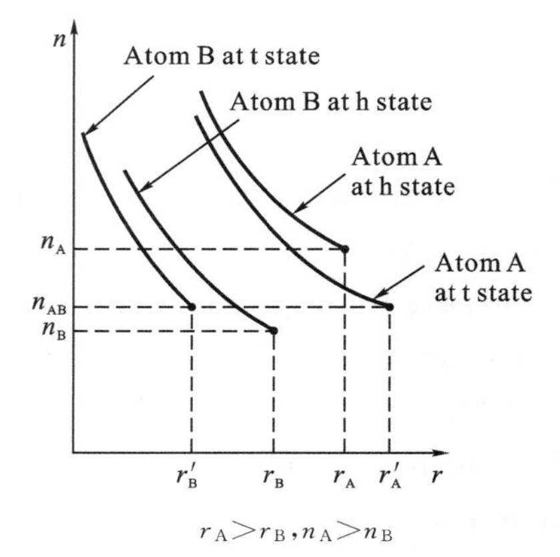

Figure 1.20 Electron density continuity between atoms.

As seen in Figure 1.19, in h state, the surface electron density of atom A is n_A and the surface electron density of atom B is n_B; in t state, the surface electron density of atom A is n_{AB} and the surface electron density of atom B is also n_{AB}. As discussed before, t state is the state when atom A and B are in contact and the electron density at the interface should be equal. Thus n_{AB} is the electron density at the interface between atom A and atom B, and it may satisfy the continuity requirement of electron density at the interface and can be referred to as the equilibrium electron density of component A and component B. For the situation in Figure 1.19, element A has smaller atomic radius and higher electron density, element B has larger atomic radius and lower electron density. For the situation in Figure 1.20, element A has larger atomic radius and higher electron density, element B has smaller atomic radius and lower electron density. The two situations correspond to different kinds of atomic phase diagrams which will be discussed in more detail later.

1.2.4 Atomic phase diagram

After two heterogeneous atoms came into contact and the valence shell electrons are rearranged, the atoms reside in a new state, the t state of the two atoms. There are two parameters describing the t state: the equilibrium electron density at the interface of the two atoms and the atomic radii of the two atoms.

If all of the solid state elements in the Periodic Table are paired up in twos, there are two cases according to atomic radius and electron density.

$$r_A < r_B, \quad n_A > n_B \tag{1-7}$$

$$r_A > r_B, \quad n_A > n_B \tag{1-8}$$

For simplicity's sake, the case corresponding to Equation (1-7) is referred to as normal combination and corresponds to Figure 1.19. The reason is that element A has smaller atomic radius and higher electron density while element B has larger atomic radius and lower electron density. The case corresponding to Equation (1-8) is referred to as abnormal combination and corresponds to Figure 1.20. The reason is that element A has larger atomic radius and higher electron density while element B has smaller radius and lower electron density. The two different kinds of combination result in two kinds of electron density versus atomic radius graph.

As shown in Figure 1.21, (a) is the electron density versus atomic radius graph corresponding to the normal combination and is able to describe the atomic state of component before and after the formation of a solid solution or a compound using atomic radius and electron density. In the graph, the horizontal axis is atomic radius, the vertical axis is electron density, point A represents component A, point B represents component B, $\bar{r}$ represents the average atomic radius of the two atoms, and $\bar{n}$ represents the average electron density of the two atoms. r_P represents the equilibrium radius of the two atoms, the electron density corresponding to r_P is n_P representing the

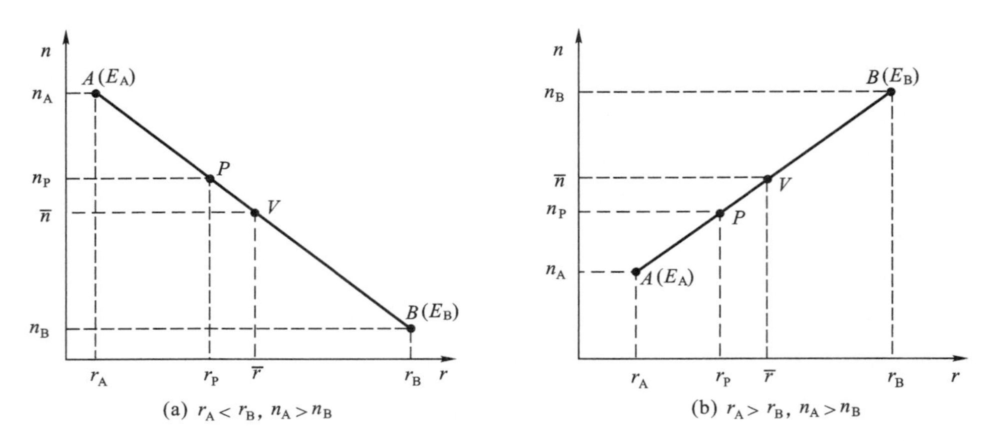

Figure 1.21 Atomic phase diagram.

equilibrium electron density. Point V and point P are called the average point and equilibrium point of atomic phase diagram respectively. The average point and equilibrium point have different physical meanings for solid solution and compounds, which is going to be discussed in relation to specific problems in another article. Figure 1.21(b) is an atomic phase diagram corresponding to the abnormal combination.

With the restriction of continuous electron density at the interface of atoms, the surface electron densities of the atoms of component A and the atoms of component B are equal in a solid solution or a AB compound. This value is the n_P in atomic phase diagram called the electron density of a solid solution or a compound. For example, when discussing the interface between compound and matrix, the surface electron density of the compound is n_P. In order to solve n_P, a formula called the lever law is needed.

1.2.5 Lever rule of atomic phase diagram

Using lever rule for classic alloy phase diagram, the relative abundance of the equilibrium phase is obtained. Using lever rule for atomic phase diagram, the interface equilibrium electron density at equilibrium state is obtained. The lever diagram of the electron density at equilibrium can be obtained from atomic phase diagram, as shown in Figure 1.22. When the atom of component A and the atom of component B reach the equilibrium electron density, the pivot point of the lever system is n_P, which is the electron density of the solid solution or compound. The two ends of the lever are n_A and n_B, which are the electron density of the atom of component A and the atom of component B respectively. The two "weights" hanging at the ends are the cohesive energy of the crystals of components A and B.

When the lever is balanced

$$(n_A - n_P)E_A = (n_P - n_B)E_B \tag{1-9}$$

Algebraic manipulation of the above expression gives

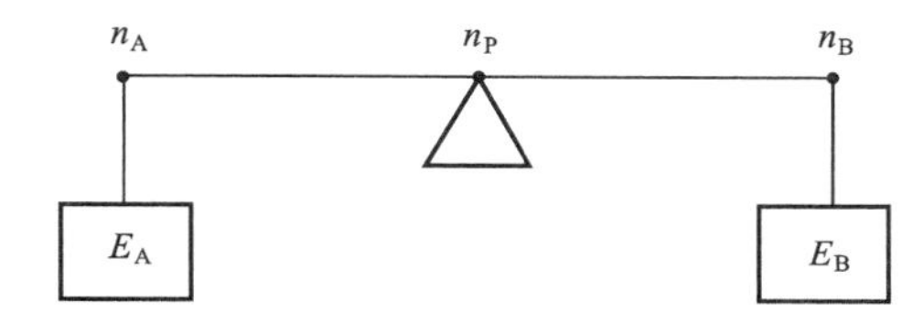

Figure 1.22 Lever rule of atomic phase diagram.

$$n_P = \frac{n_A E_A + n_B E_B}{E_A + E_B} \tag{1-10}$$

Equation (1-10) is the lever rule of atomic phase diagram. In fact Equation (1-9) also shows that the atomic radius of element with high cohesive energy are more difficult to change, thus if the cohesive energies of two components are different, in order to maintain the electron density continuity at the interface, the magnitudes of the changes of the atomic radii are different. Only when $E_A = E_B, n_P = \frac{n_A + n_B}{2} = \bar{n}$, is an approximate average electron density.

1.3 Atomic phase diagram and its application

In simple words, atomic phase diagram tries to describe how heterogeneous atoms affect each other, or how heterogeneous atoms interact with each other[16]. The first background motive for the creation of atomic phase diagram is the attempt to correct Vegard's Law[3]. In the reverse direction, that is using atomic phase diagram to explain why Vegard's Law deviates from experimental results and this will be the first application of atomic phase diagram.

Vegard's Law deviates from experimental results in most cases. In other words, Vegard's Law does not hold most of the time, but Vegard's Law is written into different kinds of textbooks and is possibly the only theory that is widely accepted but does not hold most of the time.

Since when Vegard's Law was published in 1921, people have been making a lot of efforts to correct Vegard's Law, but the problem is never solved. Actually, in order to predict how Vegard's Law deviates from experimental results, there are two problems. The first is the sign of deviation, which is either positive or negative, and the second is the amount of the deviation. If we can solve the first problem, predict the sign of the deviation, it is safe to say we almost solved the problem of Vegard's Law.

Here the idea of solving the problem of Vegard's Law is very simple: after components form a solid solution, the atomic radii of components will change in order to maintain the electron density continuity at the interface of the atoms. Atoms with higher electron density will increase its radii and atoms of lower electron density will decrease its radii. But the process of expansion and compression of atoms will not occur equally to the two kinds of atoms due to the differences in atomic radii, electron densities, and cohesive energies.

Calculations of atomic radius (i.e., lattice constant) based on atomic phase diagram of 35 unlimited solid solutions yield 28 results agreeing with experimental results. Only 7 of the calculated results does not match experimental results (80% success rate).

1.3.1 Vegard's Law

In 1921, during Vegard's investigation of ionic crystals, he discovered the linear relationship between lattice constant and composition of solid solutions, or Vegard's Law[17-18].

$$a = c_A a_A + c_B a_B \tag{1-11}$$

where a_A and a_B are the lattice constants of the solute and solvent respectively, c_A and c_B are their atomic percent respectively.

Afterward, Vegard's Law was applied to the solid solution formed by two metals with the same crystal structure. For the solid solution formed by metals of different structures, the lattice constant of each of the component need to be converted to atomic radii and Vegard's Law still applies. Vegard's Law gives a simple linear relationship between the change in composition of the components and lattice constant of solid solutions, but unfortunately Vegard's Law does not agree with experimental results in most cases. But people often say that experimental results deviate from Vegard's Law.

Crystal cell of solid solutions includes two kinds of atoms, atoms of type A and atoms of type B, and the atomic radii of the two kinds of atoms are not necessarily the same. But solid solutions have definite crystal structures and equivalent atomic radii calculated form its lattice constant, and this can be called atomic radius of a solid solution. If r_V represents the atomic radius of the solid solution calculated from Vegard's Law and r_S represent the experimental atomic radius of the solid solution, then $(r_S - r_V)$ represents the difference between experimental value and calculated value using Vegard's Law. There are two cases:

When $(r_S - r_V) > 0$, this is called positive deviation and it corresponds to the curve *MCN* in Figure 1.23. In this case, the actual lattice constant of the solid solution is greater than the calculated value using Vegard's Law.

When $(r_S - r_V) < 0$, this is called negative deviation and it corresponds to the curve *MDN* in Figure 1.23. In this case, the actual lattice constant of the solid solution is smaller than the calculated value using Vegard's Law.

People always have tried to correct Vegard's Law using elastic mechanics in order to predict the sign of the expression $(r_S - r_V)$. But the problem is never solved, the best model gives a 65% success rate predicting of the sign of $(r_S - r_V)$.

Vegard's Law only holds in rare cases, cases in which it fails are far more common. Thus Vegard's Law is the only widely accepted and applied scientific law that does not hold most of the time; this alone could prove the importance of Vegard's Law.

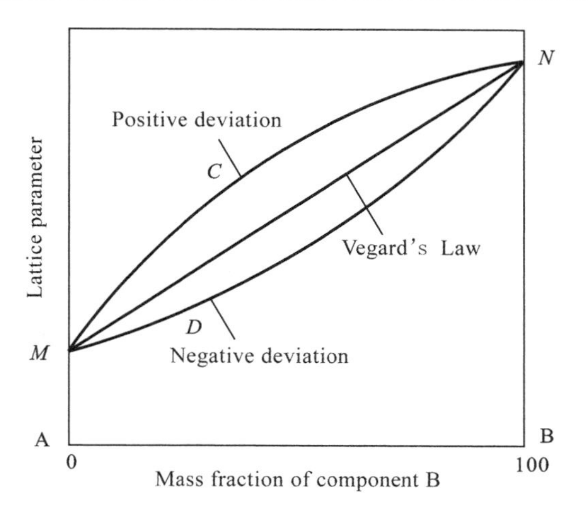

Figure 1.23 Lattice constant of solid solution vs. composition.

1.3.2 Researches about Vegard's Law in literatures

If atoms are perfectly rigid, the lattice constant of solid solutions will vary linearly with the composition of solute thus satisfying Vegard's Law. In reality, atoms are not rigid and the size of atoms can increase or decrease in solid solutions. Literatures concerning the correction of Vegard's Law often use elastic mechanics method[19-28], which does not involve the problem of atomic structure. Some typical literatures are quoted below.

In 1940, based on elastic sphere model, Pines[19] got

$$y=c_A c_B d_B - d_A\left[\frac{4\mu_A(\chi_A-\chi_B)/3}{1+4\mu_A(c_B\chi_A+c_A\chi_B)/3}\right] \tag{1-12}$$

where y is the sign of (r_S-r_V), subscript A represents solvent, subscript B represents solute, c is the concentration of atoms, d is the lattice parameters, μ is the shear modulus, χ is the compression ratio.

In 1955, based on the same model, Friedel[20] got

$$y=c_B(d_B-d_A)\left(\frac{\chi_A}{\chi_B}-1\right)\Big/\left[\frac{(1+\sigma_A)\chi_A}{2(1-2\sigma_A)\chi_B}+1\right] \tag{1-13}$$

where σ is Poisson's ratio, other symbols are as above.

One thing to point out is that Equations (1-12) and (1-13) only applies to dilute solid solutions.

In 1962, Gschneidner and Vineyard[23] also gave a formula predicting the deviation based on second-order elasticity.

$$y=2\left(\frac{d\mu}{dp}-\frac{\mu}{\kappa}\right)\frac{(d_A-d_B)^2}{d_A}c_B \tag{1-14}$$

where κ is the bulk elastic modulus of the solvent, p is the pressure.

The equation also made the approximation $c_B \ll 1$, thus only applies to dilute solution. It is obvious that the sign of y mainly depends on $\left(\frac{\mathrm{d}\mu}{\mathrm{d}p}-\frac{\mu}{\kappa}\right)$, but $\frac{\mathrm{d}\mu}{\mathrm{d}p}$ is not easily obtained.

In 2003, Lubarda[28] derived the following equation

$$y=\left[\left(1-\frac{\gamma}{\gamma_2}\right)d_A-\left(1-\xi\frac{\gamma}{\gamma_2}d_B\right)\right]c_B \tag{1-15}$$

where $\frac{\gamma}{\gamma_2}=\frac{1+4\mu_B/(3\kappa)}{1+4\mu_B/(3\kappa_A)}$, in which μ_B is the shear modulus of component B, κ_A is the bulk modulus of component A, κ is the bulk modulus of the solid solution, and ξ is a parameter related to the elasticity of the two components.

It seems that it is a pretty complicated equation and it is required to know some parameters of the solid solution.

1.3.3 Three points correct of Vegard's Law

Let the lattice constant of component A's crystal be a_A and lattice constant of component B's crystal be a_B. If component A and component B can form a unlimited solid solution, then the expression for Vegard's Law is

$$a_V=(1-c_B)a_A+c_Ba_B \tag{1-16}$$

where c_B is the percent by number of the atoms of component B in the solid solution.

According to Equation (1-16), the value of a_V can be found when $c_B=0$, $c_B=0.5$, and $c_B=1$, which is

$$a_V=\begin{cases}a_A, & c_B=0\\ \bar{a}, & c_B=0.5\\ a_B, & c_B=1\end{cases} \tag{1-17}$$

A simple case of the experiment deviating from linear relationship is the parabola. As shown in Figure 1.24, the general expression is

$$a=\alpha c_B^2+\beta c_B+\gamma \tag{1-18}$$

Figure 1.24 shows two points: the mean point $\bar{a}$, on the straight line of Vegard's Law as equation (1-16) describes; equilibrium point a_P, which should be on the experimental curve's parabola (figure shows it below the straight line of Vegard's Law), and this point will be described by Equation (1-18).

Because three points define a parabola, the only task is to find equilibrium point a_P, points a_A and a_B are known. Because lattice constant can be represented by atomic radius, thus equilibrium point a_P can be solved using atomic phase diagram. After simple algebraic derivation the following is obtained.

$$a=(c_B-1)(2c_B-1)a_A+c_B(2c_B-1)a_B-4c_B(c_B-1)a_P \tag{1-19}$$

Expression for the experimental deviation from straight line (Vegard's Law) is

$$y=a-a_V \tag{1-20}$$

In fact, to decide the sign of the deviation of Vegard's Law from experiment, it is

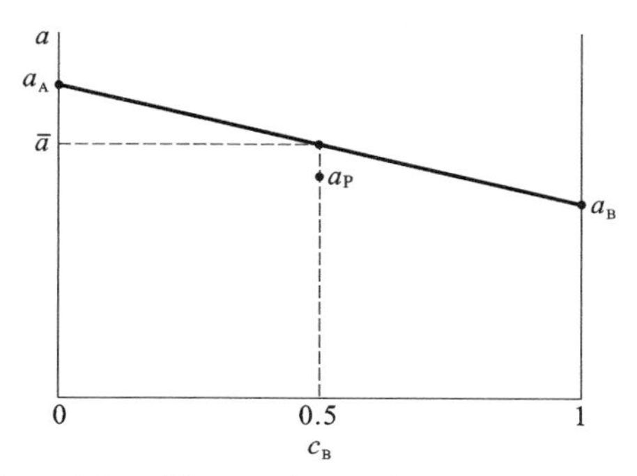

Figure 1.24 Mean point and equilibrium point.

only required to calculated the value of a_P and compare it to the experimental value. For cubic crystal, the relationship between lattice constant and Wigner-Seitz atomic radius is

$$N \cdot \frac{4}{3}\pi r_P^3 = a_P^3 \tag{1-21}$$

where N is the number of atoms in the crystal cell, r_P is the equilibrium atomic radius of the solid solution.

Convert the data in Figure 1.24 into atomic radii and label them in atomic phase diagram, then it is the situation in Figure 1.25.

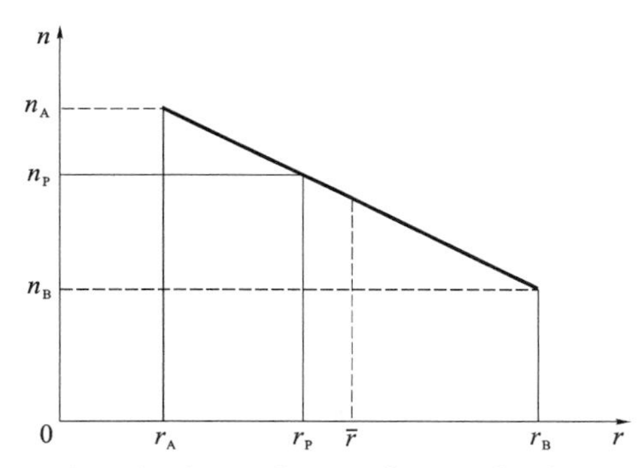

Figure 1.25 Atomic phase diagram's modify of Vegard's Law.

Perform the calculations according to Equations (1-10), (1-19), (1-20) and (1-21) to 35 unlimited solid solutions and summarize the results in Table 1.2. In the table, r_V is calculated value using Vegard's Law, r_P is the calculated value using atomic phase diagram, $r_P - r_V$ is the predicted sign of deviation of Vegard's Law by atomic phase diagram, $r_{exp} - r_V$ is the sign of experimental deviation from Vegard's Law, the units of atomic radius and electron density are the same as in Table 1.1. From the table one can see that the atomic phase diagram predictions for only 7 alloy systems, namely, Nb-Mo, Mo-V, Mo-Ta, V-W, Ni-Au, Ni-Pd and Pt-Pd, does not agree with the experimental value. One thing to point out is that in this theory only three physical quantities, namely, atomic radius, electron density, and cohesive energy for the component crystal, are needed.

Table 1.2　Vegard's Law corrected by atomic phase diagram

System	r_V	r_P	$r_P - r_V$	$r_{exp} - r_V$
Ag-Pd	1.5603	1.5550	−0.0053	<0
Ag-Au	1.5957	1.5954	−0.0003	<0
Au-Pt	1.5634	1.5571	−0.0063	<0
Au-Cu	1.5026	1.5067	0.0041	>0
Au-Pd	1.5578	1.5574	−0.0004	<0
Cu-Ni	1.3950	1.3929	−0.0021	<0
Cu-Pd	1.4672	1.4703	0.0031	>0
Cu-Rh	1.4496	1.4588	0.0092	>0
Cu-Pt	1.4728	1.4882	0.0154	>0
Cr-Fe	1.4161	1.4160	−0.0001	<0
Cr-Mo	1.4853	1.5016	0.0163	>0
Bi-Sb	1.9878	1.9822	−0.0056	<0
Nb-W	1.5905	1.5878	−0.0027	<0
Nb-Zr	1.6992	1.6920	−0.0072	<0
Nb-Mo	1.5871	1.5890	0.0019	<0
Nb-V	1.5569	1.5686	0.0117	>0
Mo-Ti	1.5820	1.5767	−0.0053	<0
Mo-W	1.5539	1.5543	0.0004	>0
Mo-V	1.5203	1.5240	0.0037	<0
Mo-Ta	1.5896	1.5930	0.0034	<0
V-Ti	1.5519	1.5491	−0.0028	<0
V-Ta	1.5595	1.5739	0.0144	>0
V-W	1.5237	1.5322	0.0085	<0
Ir-Pd	1.5120	1.5090	−0.0030	<0
Ir-Pt	1.5176	1.5162	−0.0014	<0
W-Ta	1.5931	1.5914	−0.0017	<0
W-Cr	1.4888	1.5141	0.0253	>0
Ni-Au	1.4856	1.4775	−0.0081	>0
Ni-Pd	1.4502	1.4455	−0.0047	>0
Ni-Pt	1.4558	1.4665	0.0107	>0
Ni-Rh	1.4327	1.4397	0.0070	>0
Pb-Tl	1.9161	1.9168	0.0007	>0
Pt-Rh	1.5105	1.5107	0.0002	>0
Pt-Co	1.4591	1.4697	0.0106	>0
Pt-Pd	1.5280	1.5291	0.0011	<0

1.4 Summary

Although this is the first chapter of the book, it would be wise to look over "Research on superplasticity of Zn-Al eutectic alloys" in later sections especially chapter 4 first in detail.

Zn-5Al alloy displays a 5000% elongation, such property is mainly due to the abundance of Zn/Al interfaces in the alloy. Diffusion and dissolution process occurs at Zn/Al interface at 350 ℃ which results in a diffusion dissolution zone. Superplastic deformation occurs right inside the diffusion dissolution zone.

After any two alloy phase came into contact at a temperature high enough for atoms to diffuse, a diffusion dissolution zone will form at the interface where the two in contact. Diffusion dissolution zone in interface can contain many phases, these phases can be either solid solutions or intermetallic compounds.

In order to describe diffusion dissolution zone, a new idea, atomic phase diagram, is proposed. Maybe atomic phase diagram will develop into a theory, but under current circumstances atomic phase diagram is only an unborn child. The atomic phase diagram described in this chapter is only a kind of restlessness inside the mother's womb.

Atomic phase diagram does not solve the problem of Vegard's Law deviating from experimental results, but it provides some kinds of hope.

Appendix B provides the equilibrium electron density data after contact with heterogeneous atoms of common metals, which is an important parameter for atomic phase diagram. It is hoped that readers will be able to find new uses for these data.

Chapter 2

Superplasticity of Zn-Al Eutectic Alloy

In this chapter, first the experimental methods of superplastic tensile test are introduced. Then the superplastic tensile test results are summarized. The superplastic deformation characteristics of Zn-5Al eutectic alloys are discussed in detail.

For Zn-Al alloys, a review about the superplasticity was written by Underwood in 1962[29], which indicates that the Zn-22Al eutectoid alloy has a maximum elongation of 650%, as shown in Figure 2.1. Today, the Zn-5Al eutectic alloy has the elongation of 5000%.

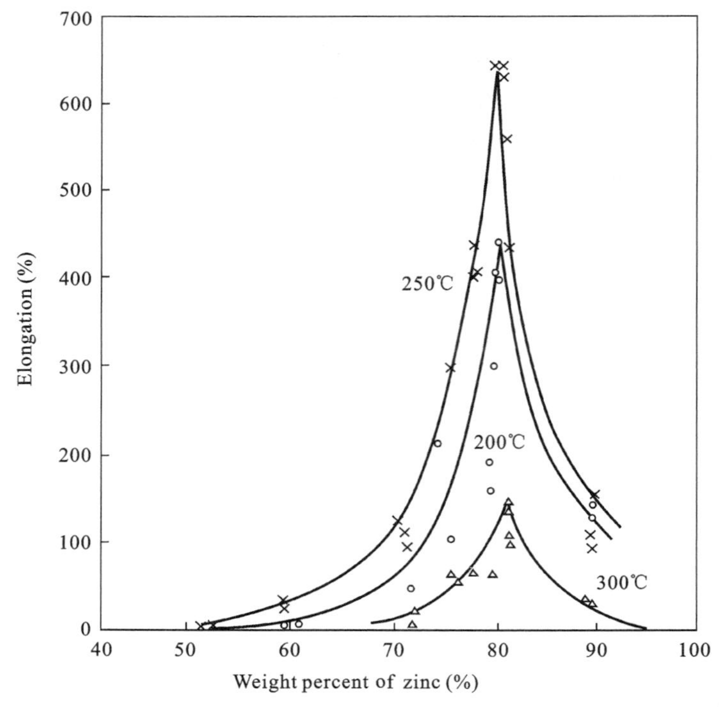

Figure 2.1 The superplasticity in Zn-Al alloys reviewed by Underwood[29].

Zn-22Al alloy is an eutectoid alloy whose superplasticity can be obtained by means of homogeneous annealing at 365±15 ℃ for 50 hours, deformation at 260±10 ℃ by

60%, solid solution at 365±15 ℃ for 5 hours and quenching in water, deformation at range of 300±25 ℃ by 80%, aging at 225±25 ℃ for 1 hour. All these are called pre-processing for superplasticity.

Superplasticity of Zn-5Al eutectic alloys is superior to that of Zn-22Al and its superplastic pre-processing is also much simpler. By direct rolling after the ingots cooling to 250±50 ℃, and then annealing at 350±5 ℃ for 2 hours followed rolling again, thus obtained sheets can exhibit excellent superplasticity.

2.1 Experimental methods

The superplastic tensile experiments are of three types: constant velocity, constant load and constant stress. All the equipment in these three methods is designed and manufactured by the author. The self-made equipment is simple, practical and cheap. The design principles, the experimental accuracy and the recorded data reliability of the equipment are described briefly as follows.

A self-made machine with constant speed is shown in Figure 2.2. It can record the change of load with time, viz., the P vs. t curves, under the condition of constant speed. The speed of the machine has 15 levels from 10^{-5} mm/s to 10^{-1} mm/s and its temperature accuracy is ±2 ℃. The elongation of 5000% in Zn-5Al alloy was obtained by using this machine, as shown in Figure 2.3. The gauge of the specimen developed from 10 mm to 510 mm and the specimen was not broken yet.

Figure 2.2 The tensile test machine with constant speed.

Compared with constant speed method, the equipment required for constant load method is much simpler. The variation of displacement with time, viz., the Δx vs. t curves, can be obtained. The accuracy of temperature, load, time and displacement is ±5 ℃, ±0.1 g, ±0.5 s and 0.5 mm, respectively.

The equipment for constant stress method may be reconstructed on the basis of

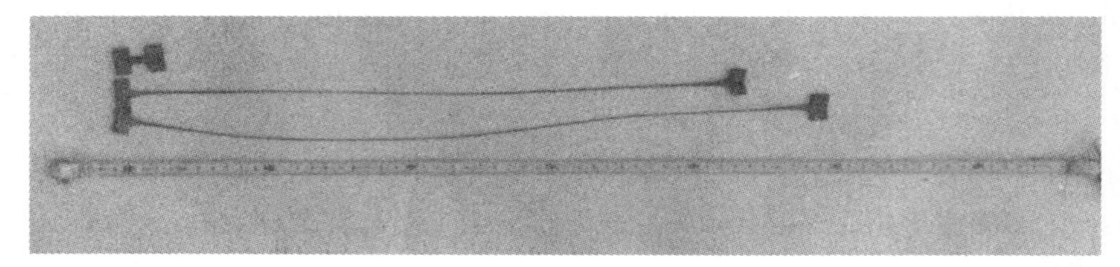

$T=350$ ℃, $\dot{\varepsilon}=5.4\times10^{-4}$/s

Figure 2.3 The elongation of 5000% for Zn-5Al superplastic alloy.

that for constant load. During the tensile deformation, due to the continuous decreasing of cross section area of specimen, the load should be correspondingly decreased to keep constant stress.

2.1.1 Deformation characteristic in Zn-5Al alloy

The original data obtained from constant load method is the relationship of displacement and time, i.e., the Δx versus t curve. In the experiments, the gauge length of the specimen x is the function of time t, i.e.

$$x=f(t) \tag{2-1}$$

Instant strain at time t can be written as

$$\mathrm{d}\varepsilon=\frac{\mathrm{d}x}{x} \tag{2-2}$$

The total strain produced from the beginning of the test to the time t can be represented as

$$\varepsilon(t)=\int_{x_0}^{x(t)}\frac{\mathrm{d}x}{x}=\ln\left[1+\frac{\Delta x(t)}{x_0}\right] \tag{2-3}$$

where $\Delta x(t)$ is the displacement, namely, the absolute extension of specimen.

Figure 2.4 is the characteristic Δx versus t curve for Zn-5Al alloys during superplastic tensile test.

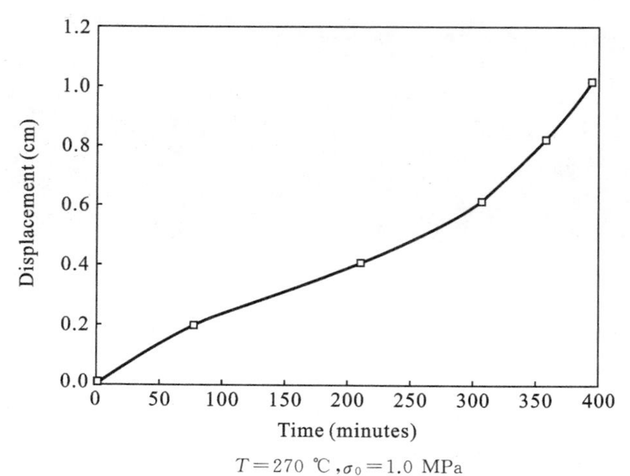

$T=270$ ℃, $\sigma_0=1.0$ MPa

Figure 2.4 The Δx versus t curve for Zn-5Al alloy.

The typical ε versus t creep curve for Zn-5Al alloy is shown in Figure 2.5. The curve can be divided into three zones: primary state zone, quasi-steady state zone and accelerating state zone. In this chapter, the emphasis will be placed on the quasi-steady state zone. As shown in Figure 2.6, so-called quasi-steady creep is that strain rate is approximately a constant in the QSS zone.

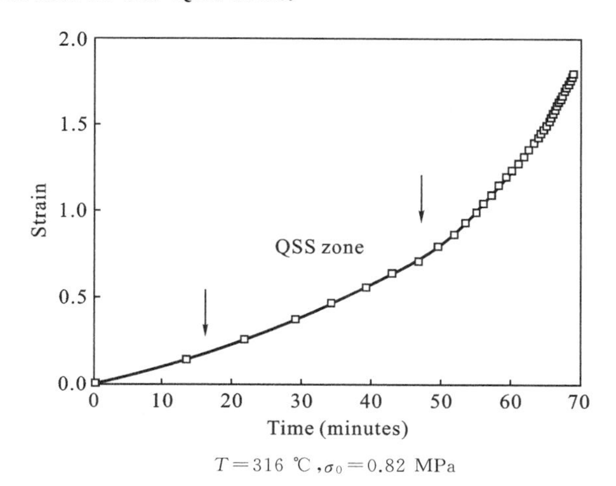

$T=316$ ℃, $\sigma_0=0.82$ MPa

Figure 2.5 The ε versus t curve for Zn-5Al alloy.

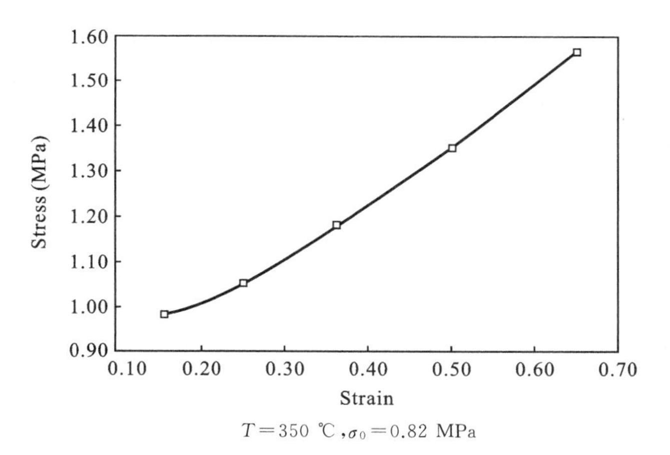

$T=350$ ℃, $\sigma_0=0.82$ MPa

Figure 2.6 The σ versus ε curve for Zn-5Al alloy.

2.1.2 Rheology of Zn-5Al alloys

Superplastic deformation behaves as a viscous flow when reaching steady state. Figure 2.7 shows the tensile results of Zn-5Al alloy at the temperature of 350 ℃ and the initial stress of 0.82 MPa. Figure 2.7 is the $\dot{\varepsilon}$ versus σ curve corresponding to Figure 2.6. It is seen that σ has the linear relationship with $\dot{\varepsilon}$, i.e.

$$\sigma=\eta\dot{\varepsilon} \tag{2-4}$$

where η is the viscous coefficient.

According to Equation (2-4), the viscous coefficient of steady-state flow in Zn-5Al

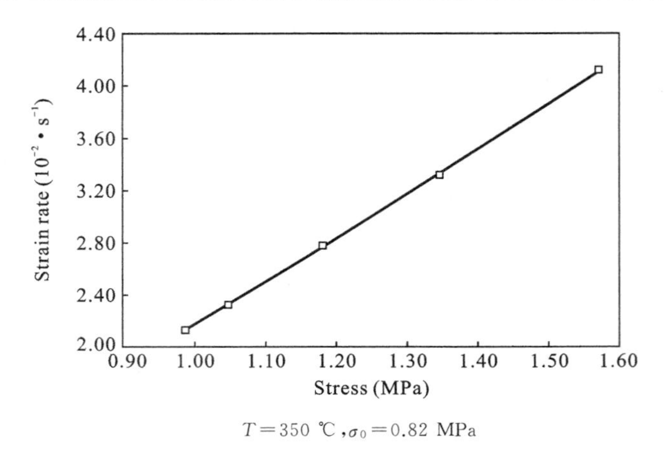

$T=350$ ℃, $\sigma_0=0.82$ MPa

Figure 2.7 The $\dot{\varepsilon}$ versus σ curve for Zn-5Al alloy.

alloy at the temperature of 350 ℃ as shown in Figure 2.7, η is 2000 MPa · s, and the corresponding strain rate sensitivity exponent m is below 0.4.

The strain rate sensitivity exponent m was proposed by Nadai and Manjoine[30] in as early as 1941. In 1964, Backofen and Avery[31] applied the m value to superplasticity.

Backofen and Avery defined the strain rate sensitivity exponent by

$$m_x=\left(\frac{\mathrm{dln}\,\sigma}{\mathrm{dln}\,\dot{\varepsilon}}\right)_x=\frac{\Delta\ln P}{\Delta\ln v}=\frac{\ln(P_B/P_A)}{\ln(v_B/v_A)} \tag{2-5}$$

where v_A and P_A are the velocity and the load of deformation in state A, and v_B and P_B are the corresponding values in state B, which can be measured by means of speed leap in the tensile test.

Song Yuquan[32] defined the strain rate sensitivity exponent by

$$m_p=\left(\frac{\mathrm{dln}\,\sigma}{\mathrm{dln}\,\dot{\varepsilon}}\right)_p=\frac{1}{\Delta\ln v/\Delta\ln x-1}=\frac{1}{\ln(v_B/v_A)/\ln(x_B/x_A)-1} \tag{2-6}$$

where x_A, v_A, x_B and v_B are the displacement and the velocity corresponding to state A and state B respectively, which can be obtained from the Δx versus t curve.

It can be followed from Equations (2-5) and (2-6) that the m value comes from the change of two states.

Figure 2.8 is a plot of the tensile test curve for Zn-5Al alloy at the temperature of 316 ℃ and under the constant load and the initial stress of 0.82 MPa. The m_p value in the figure is the strain rate sensitivity exponent obtained from Equation (2-6). The m_p value is sensitively related to the strain rate. In the case of higher strain rate, the m_p value is larger and the corresponding viscous coefficient is smaller. Figure 2.9 is the σ versus ε curve corresponding to the zone, $m_p>0.9$ in the Figure 2.8 and the corresponding $\dot{\varepsilon}$ versus σ curve is shown in Figure 2.10. When $m_p>0.9$, as shown in Figure 2.10, the viscous coefficient of superplastic flow deduced from Equation (2-4) is 350 MPa · s; when $m_p<0.4$, the viscous coefficient is 3000 MPa · s.

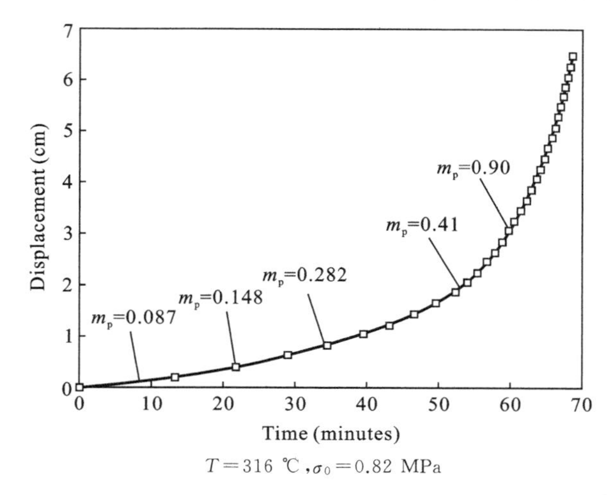

$T=316$ ℃, $\sigma_0=0.82$ MPa

Figure 2.8 Tensile test under constant load and its m value.

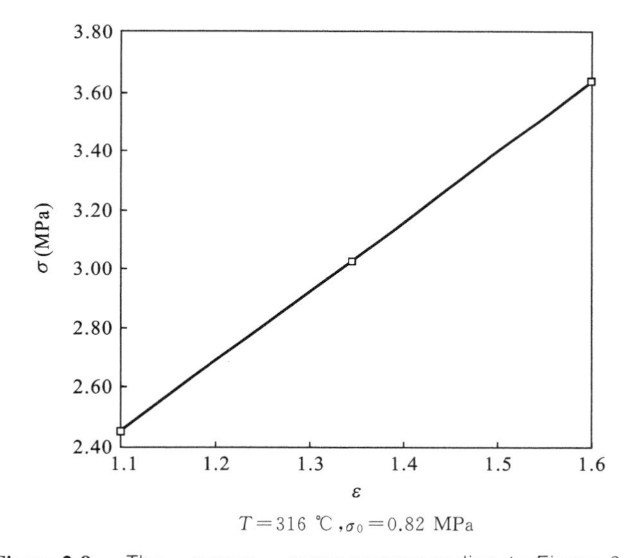

$T=316$ ℃, $\sigma_0=0.82$ MPa

Figure 2.9 The σ versus ε curve corresponding to Figure 2.8.

2.2 Factors affecting superplasticity in Zn-5Al alloy

The superplasticity in Zn-Al alloy can be affected by the amount of Al, the annealing time and the storing time, and so on.

2.2.1 Effect of Al on superplasticity in Zn-Al alloy

Figure 2.11 shows the elongation of superplastic tensile test for Zn-5Al, Zn-2.5Al and Zn-10Al alloys at the same temperature and deformation speed, among which the elongation of the Zn-5Al eutectic alloy is the largest.

It is the Zn-5Al eutectic alloy that has the smallest deformation stress among three

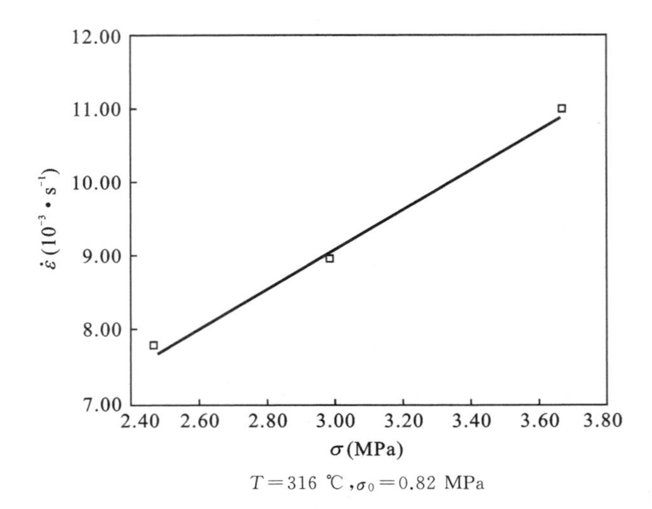

Figure 2.10 The $\dot{\varepsilon}$ versus σ curve corresponding to Figure 2.9.

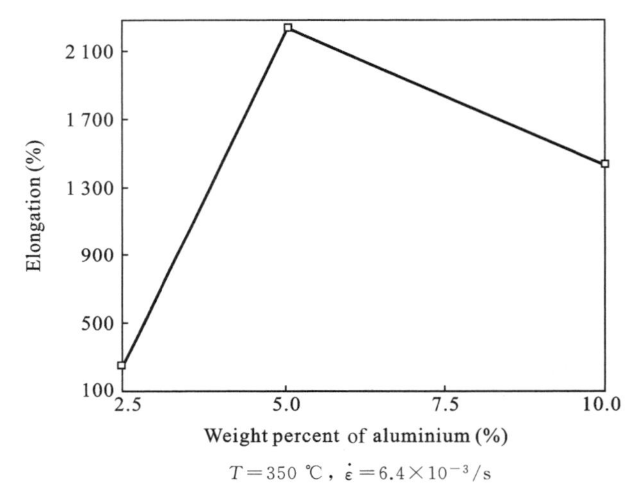

Figure 2.11 Effect of Al on elongation of Zn-Al alloys.

alloys at the same deformation temperature and speed, as shown in Figure 2.12.

2.2.2 Effect of annealing time on superplasticity in Zn-5Al alloy

The annealing of specimen before superplastic deformation has obviously effect on the elongation of superplastic deformation, as shown in Figure 2.13. The effect of annealing time on superplastic deformation stress is shown in Figure 2.14. The time reaching the set deformation temperature for the specimen is generally ten minutes, and further annealing before the beginning of superplastic tensile test will decrease the elongation of the alloys and increase the stress of deformation. The effect of annealing time on superplasticity has a critical point, about 5 hours. The annealing time less than the critical point significantly affects the superplasticity of specimen and the annealing time more than the critical point does not have pronounced effect on the superplasticity.

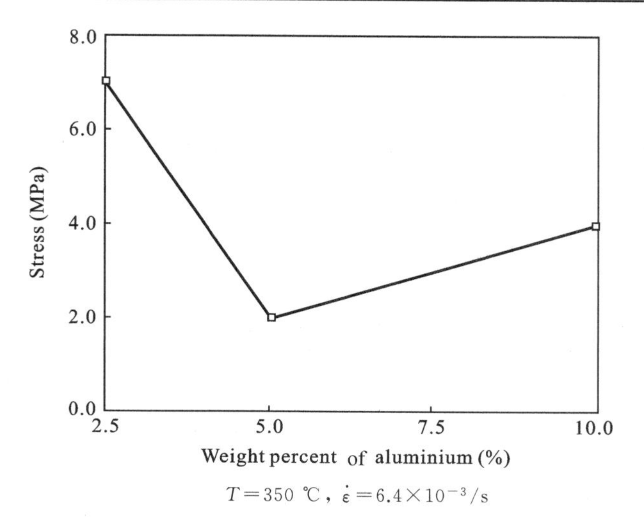

$T=350$ ℃, $\dot{\varepsilon}=6.4\times10^{-3}$/s

Figure 2.12　Effect of Al on deformation stress of Zn-Al alloys.

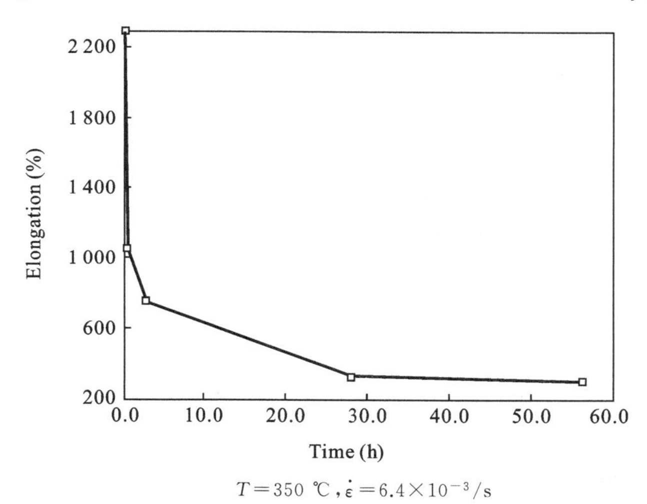

$T=350$ ℃, $\dot{\varepsilon}=6.4\times10^{-3}$/s

Figure 2.13　Effect of annealing time on elongation of Zn-5Al alloy.

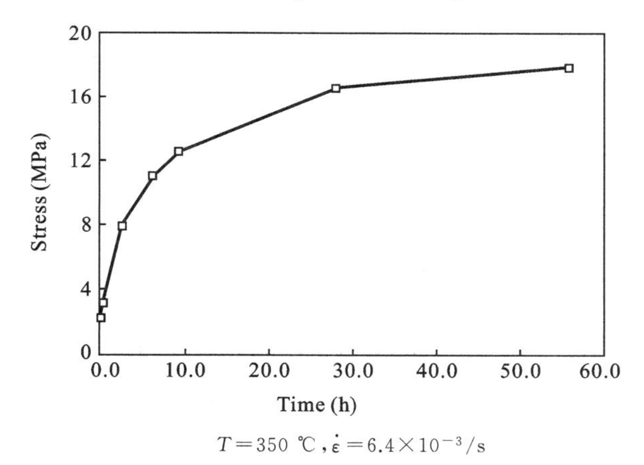

$T=350$ ℃, $\dot{\varepsilon}=6.4\times10^{-3}$/s

Figure 2.14　Effect of annealing time on deformation stress of Zn-5Al alloy.

The beginning of annealing corresponds to the recovery and recrystallization in the specimen.

2.2.3 Effect of natural aging on superplasticity in Zn-5Al alloy

Table 2.1 shows the natural aging characteristics of superplastic Zn-5Al alloy. The elongation of the tensile test decreases with the increasing of the time of alloys being aged at room temperature.

Table 2.1 Effect of natural aging on superplasticity in Zn-5Al alloy

Aging(year)	1	2	3	4	5	6	7
Elongation(%)	4400	4400	3600	3500	2900	2400	2400

2.3 Energy analysis of superplasticity

Zn-5Al alloy in rolled state exhibits low stress and large elongation during superplastic deformation. Certain energy will be stored within the alloys after cold rolled deformation, according to the viewpoint of thermodynamics. In the subsequent heat treatment, the stored energy will be released. Annealing treatment prior to superplastic deformation is the course of releasing stored energy. For Zn-5Al alloy in rolled state, the superplastic deformation starting from the rolled state is actually accompanied by the releasing of stored energy, which is a driving force for the superplastic deformation.

After superplastic deformation by an elongation of 5000%, Zn-5Al alloy still has certain stored energy, as shown in Figure 2. 15. The photograph (a) was taken immediately after the specimen being taken out of the furnace on test machine, and the photograph (b) was taken after two years of horizontal placing when the spontaneous deformation occurred on the horizontal plane, which proves that there still exists certain energy in the specimen after superplastic deformation of 5000%.

The deformation of an alloy can be analyzed in terms of the first law of thermodynamics. For cold work on the alloys, the first law can be written as

$$\Delta E = W - Q \tag{2-7}$$

where ΔE is the change of internal energy in the alloys, Q is the heat released during cold work and W is the work done by the applied force.

Stored energy can be defined as

$$E_s \equiv \Delta E = W - Q \tag{2-8}$$

Generally, $W > Q > 0$.

Some methods of measuring the stored energy of cold work are to determine the change in internal energy ΔE. Others are to determine the change in enthalpy ΔH. At constant pressure, however

$$\Delta H = \Delta E + p\Delta V \tag{2-9}$$

where p is the hydrostatic pressure and ΔV is the volume change associated with the

(a) Deformation of the day

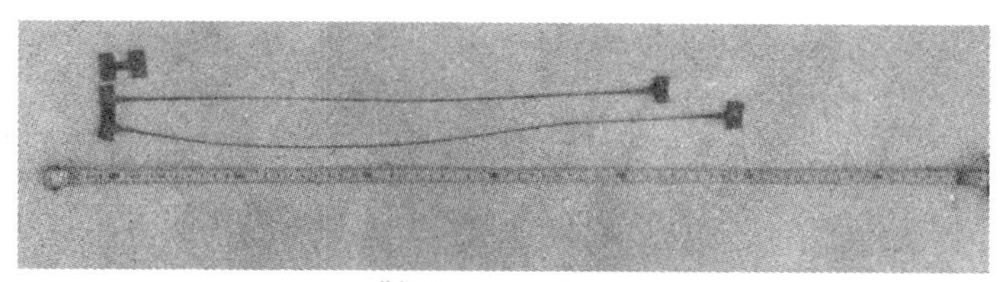

(b) Two years later

$T=350$ ℃, $\dot{\varepsilon}=5.4\times10^{-4}/\text{s}$, $\delta_{max}=5000\%$

Figure 2.15 Spontaneous deformation of Zn-5Al alloy after superplastic deformation.

process.

Since the product $p\Delta V$ is negligibly small for solids at or near atmospheric pressure, the enthalpy change ΔH is approximately equal to the change in internal energy ΔE, which is the stored energy E_s.

Generally, the stored energy in alloys cannot be increased infinitely. The more the work W used for deformation, the higher the corresponding E_s, but the ratio of E_s to W will decrease with increasing degree of deformation. This is because that, at high degree of cold work deformation, E_s is close to a constant. Hence, when deformation reaches a certain stage, the stored energy can not be significantly increased again. Such a deformation is called saturated deformation. When deformation reaches the saturated state, the work done by the applied force is completely dissipated into heat.

According to theory of elasticity and plasticity, the cold work stored energy in alloys can be represented as

$$W^r=\frac{1}{2}\int\sigma^r : C^{-1} : \sigma^r \mathrm{d}v \tag{2-10}$$

where C is a fourth rank tensor of elastic coefficient, σ^r is the residual stress in alloys, W^r is the stored energy related with the residual stresses within the materials[33].

With reference to Chapter 5, in the case $\sigma^r=90$ MPa, $C=8.97\times10^4$ MPa, for the rolled Zn-5Al alloy, stored energy deduced from the above equation, W^r is 4.52×10^4 J/m^3, i.e., 23.4 J/mol.

Stored energy of cold work in the metals is generally at the order of 50 J/mol or so. Even at large strain, stored energy is only about 800 J/mol[34]. There are many methods to measure stored energy, which can be divided into single-step methods and two-step methods in terms of operating procedures. For the former, all measurements are made during the deformation, and while the latter is that the deformation is carried out first and the stored energy is measured at a later time.

In two-step methods, the stored energy is found by comparing the thermal behavior of the cold-worked specimen with that of a standard specimen in a suitable process. In the case of metallic elements and some alloys, the standard specimen is in the annealed condition. Two-step methods are of two types: annealing methods and reaction methods. Annealing methods may be further subdivided into: anisothermal annealing, in which the temperature of the specimen is increased continuously, usually at a constant rate, and the effect of the energy release is observed as a function of temperature; isothermal annealing, in which the specimen is placed in a constant-temperature calorimeter and the effect of the energy release is observed as a function of time at a selected annealing temperature. In reaction methods, the cold-worked and standard specimens respectively react with a working substance in a calorimeter, and the energy is obtained as the difference between the heats of reaction. The anisothermal annealing method is used in this section.

A heat analysis system of Dupont-1090B is used in the experiments, and the specimens are measured by Differential Scanning Calorimeter (DSC). The specimens of Zn-5Al alloys are heated to 370 ℃ at rates of 10 ℃/min and 5 ℃/min respectively. The heating power dH/dt versus temperature T curves are obtained for the specimens in various states, and the experiments in every state are repeated three times.

2.3.1 Stored energy in rolled alloy

Figure 2. 16 is the DSC measurement result of rolled Zn-5Al alloy. In the experiment, the dose of the specimen is 10 mg, and the reference specimen is Al_2O_3. In the figure, the longitude axis is the compensating power and the transverse axis is the temperature, and the heating rate is 10 ℃/min.

According to the principle of DSC, the compensating power is given by

$$\Delta W=\frac{dQ_s}{dt}-\frac{dQ_r}{dt}=\frac{dH}{dt} \tag{2-11}$$

where Q_s is the heat absorbed by the specimen, Q_r is the heat absorbed by the reference material, dH/dt is the change of enthalpy per second, namely, heat flow rate (mJ/s).

It is seen that the negative sign of the longitude axis in Figure 2.16 comes from Equation (2-11) about definition of heat.

For the recrystallized Zn-5Al alloy by 350 ℃ × 6 h, the DSC curve is shown in Figure 2.17, and the meaning of the sign in the figure is the same as above. If the influences of the backgrounds from 40 ℃ to 370 ℃ in Figures 2. 16 and 2. 17 are neglected, it may be thought that the heat absorbed by the rolled specimen, Q_1 is 7.72 J/g, and the heat absorbed by the recrystallized specimens, Q_2 is 8.46 J/g. It is clear that the difference of the heats absorbed by these two specimens is stored energy. Thus,

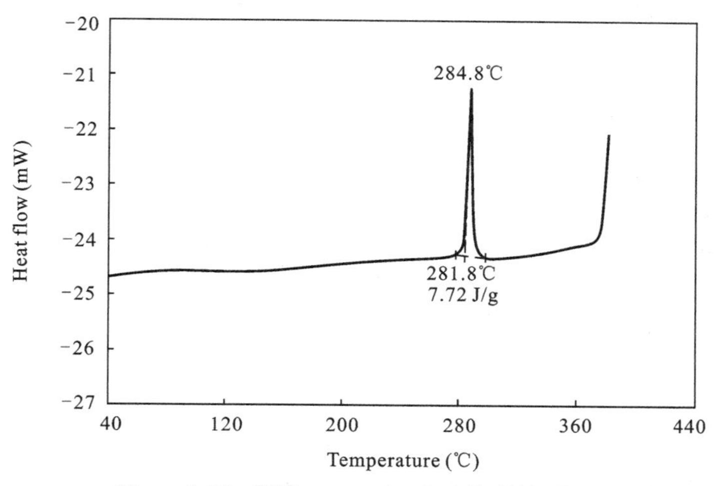

Figure 2.16 DSC curve of rolled Zn-5Al alloy.

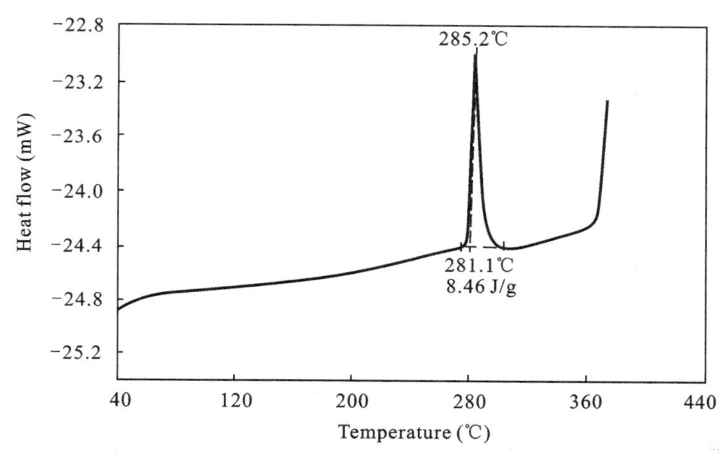

Figure 2.17 DSC curve of recrystallized Zn-5Al alloy.

by contrast with the recrystallized specimen, the stored energy in rolled specimen is

$$E_s = Q_2 - Q_1 = 48 \text{ J/mole} \tag{2-12}$$

Equation (2-12) shows that the energy of 48 J/mole is released in the rolled specimen of Zn-5Al alloy, after recrystallization by 350 ℃×6 h.

2.3.2 Stored energy in superplastically deformed alloy

For superplastically deformed specimen of Zn-5Al alloy, at the temperature of 350 ℃, the strain rate of 5.4×10^{-4}/s, and by the elongation of 2500%, its DSC curve is shown in Figure 2.18. Besides the peak at the eutectoid temperature, there exists another peak at 131 ℃, whose mechanism is still not clear. The heat absorbed by the specimen during the whole process of measurement is 9.45 J/g. According to Equation (2-12), the stored energy released by Zn-5Al alloy in superplastically deformed state by an elongation of 2500% is 112 J/mol.

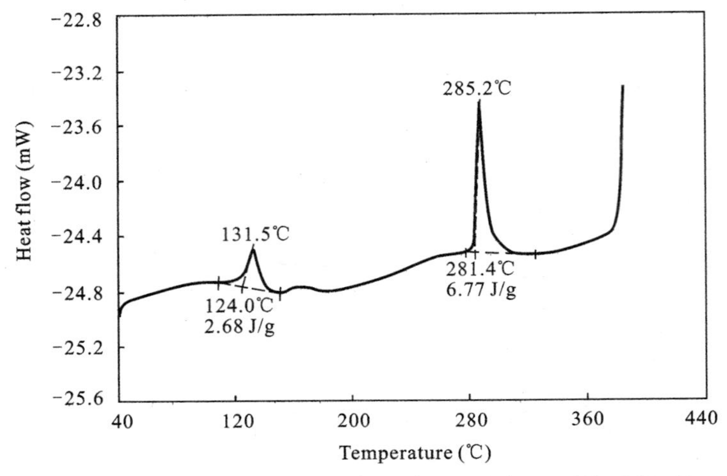

Figure 2.18 DSC curve of superplastically deformed Zn-5Al alloy.

Figures 2.19 and 2.20 respectively are the DSC curves for Zn-5Al specimens in the rolled state and the superplastically deformed state at the temperature of 350 ℃, the strain rate of 5.4×10^{-4}/s, and the elongation of 600%, with the recrystallized specimen by 350 ℃×72 h as reference. It is seen that the peaks in Figure 2.19 are just in the reverse directions of those in Figure 2. 20, which indicates that the superplastically deformed state is opposite to the rolled state.

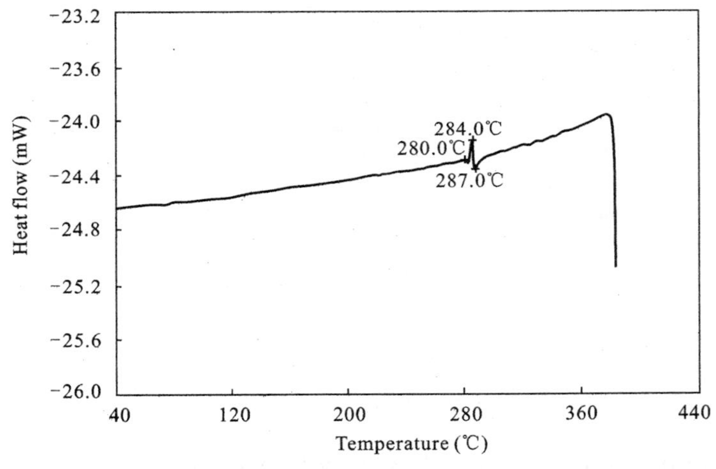

Figure 2.19 DSC curve of rolled specimen with reference to recrystallized specimen.

2.4 Release of stored energy during superplastic deformation

The energies of the system can be divided into dissipative energy and effective energy during deformation of materials[35]. A characteristic σ versus ε curve for superplastic deformation is shown in Figure 2.21. The dissipative energy density during deformation is β, i.e., the area *opq* as shown in Figure 2.21; the effective energy density is α, i.e., the area qpg as shown in the figure. α may be understood as dynamic elastic-plastic stored energy during superplastic deformation.

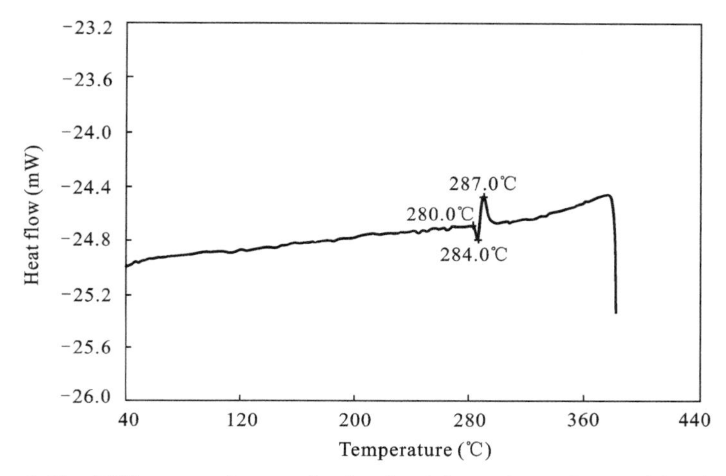

Figure 2.20 DSC curve of superplastically deformed specimen with reference to recrystallized specimen.

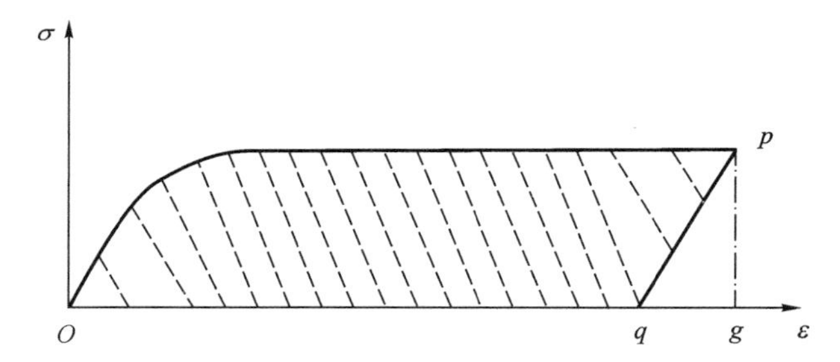

Figure 2.21 Dissipative energy and effective energy densities during deformation of materials.

2.4.1 Releasing rate of stored energy

Dissipative energy during deformation of materials can be divided into two parts: intrinsic stored energy and foreign stored energy. The former is stored energy of cold work in preparing materials, the latter is dynamic stored energy provided by applied force during deformation, and it is very small. Under the condition of largest elongation, if the stored energy is completely dissipated, then the dissipative energy in Figure 2.21 can be given by

$$\beta = E_s + E_a \tag{2-13}$$

where E_s is the stored energy of the system, E_a is the energy provided by the applied force, β is the dissipative energy during superplastic deformation.

In fact, due to extremely low stress about 1 MPa to 10 MPa and very high strain during superplastic deformation, the area opq in Figure 2.21 is extremely large and the area qpg is very small. In other words, the dissipative energy is far larger than the effective energy. According to classical mechanics, the dissipative energy is given by

$$\beta = \sigma\varepsilon \tag{2-14}$$

where σ is the stress, ε is the strain and β is the dissipative energy density.

Equations (2-13) and (2-14) are differentiated about time, if E_a does not change with time at constant stress tension, then

$$\dot{\varepsilon} = \dot{E}_s/\sigma \tag{2-15}$$

where $\dot{\varepsilon}$ is the strain rate, $\dot{E}_s$ is the dissipative rate of the stored energy and σ is the stress.

The equation shows that the strain rate depends on the releasing rate of the stored energy.

2.4.2 Anelasticity and internal friction

The effective energy can be thought as elastic strain energy according to elasticity theory, and according to Hook Law, the effective energy is the area qpg in Figure 2.21.

For the real materials, there exists a response time for loading and unloading. In so-called elastic zone, in fact, the stress does not change linearly with the strain, and this is anelasticity. The characteristic of the anelasticity is that, when loading or unloading, the stresses and the strains reach equilibrium values through a process of relaxation rather than instantly. Therefore, the relaxation can be divided into the stress relaxation and the strain relaxation.

If the elastic strain energy is transformed into heat before relaxation, then this phenomenon is called internal friction. The superplastic Zn-Al alloys have obvious internal friction effect, consequently, internal friction is also one of characteristics of superplasticity. In fact, the line qp in Figure 2.21 is not straight, consequently, the process in the system is nonlinear.

2.4.3 Energy parameter controlling superplastic deformation

It is similar to the stress induced ordered internal friction discussed by Ha Kuanfu[36] that energy parameter controlling superplastic deformation will be discussed in this paragraph. Let U is the total energy of the system, p is the degree of order, which can measure the ability of uniform deformation of materials. When the parameter p increases with one unit, under the condition of constant strain, the energy decreased in the system is given by

$$u = -\left(\frac{\partial U}{\partial p}\right)_{\varepsilon} \tag{2-16}$$

For constant temperature system

$$\mathrm{d}U = \sigma \mathrm{d}\varepsilon - u\,\mathrm{d}p \tag{2-17}$$

where σ is the stress, ε is the strain, u is the intensity of energy controlling the system, p is the degree of order about uniform superplastic deformation.

Differentiate $(U-\sigma\varepsilon)$, i.e.

$$d(U-\sigma\varepsilon) = \sigma d\varepsilon - u dp \quad (2\text{-}18)$$

Equation (2-18) is a perfect differential, then

$$\left(\frac{\partial u}{\partial \sigma}\right)_p = \left(\frac{\partial \varepsilon}{\partial p}\right)_\sigma \quad (2\text{-}19)$$

Equation (2-19) shows that during superplastic deformation under constant stress, if the change of degree of order causes the change of strain, then the applied force will increase the energy for ordering, which makes the system develop towards the state whose the degree of order is increasing. As a result, the materials can continue to deform uniformly.

2.4.4 Fluctuation and relaxation

When there is a fluctuation in a system of deformation, certain random behavior will occur within the specimen. The time required for the degree of order p to reach its equilibrium value is defined as relaxation time. If the system can make p reach its equilibrium value within relaxation time, superplastic deformation may continue, otherwise, disordered deformation will take place and then fracture will arise in superplastic deformation.

The relaxation in the deformation process of materials is atomic process on microscopic scale, therefore, relaxation time τ depends on temperature and if their relationship obeys Arrhenius equation, then

$$\tau = \tau_0 \exp[-H/(kT)] \quad (2\text{-}20)$$

where H is the activation energy in the process, τ_0 is a constant, k is the Boltzmann constant, T is the absolute temperature.

2.5 Summary

The Zn-5Al eutectic alloy can obtain an elongation of 5000% during superplastic deformation, after the alloy is prepared. With the increasing of deposit time, the elongation of the alloy decreases annually with an average amount of 8%.

Before superplastic deformation for Zn-5Al alloy, the annealing treatment toward the specimen will decrease elongation and increase stress. Within the critical point, the annealing treatment has a pronounced effect on superplasticity.

The deviation of the composition of Zn-Al alloys from the eutectic point will result in decreasing of elongation and increasing of stress during superplastic deformation. The decreasing amount of Al in the alloy will lead to more pronounced decreasing of elongation and increasing of stress.

The strain rate sensitivity exponent of superplastic deformation for Zn-5Al eutectic alloy at constant load increases with the increasing of strain rate and approximates to unity in the last fracture process.

When Zn-5Al eutectic alloy deforms at constant load, the viscous coefficient of superplastic flow is 3000 MPa • s and 350 MPa • s in the case $m<0.4$ and $m>0.9$ respectively.

The stored energy of 48 J/mole in rolled Zn-5Al alloy is released after recrystallization by 350 ℃ ×6 h. The stored energy of 112 J/mole in rolled Zn-5Al alloy is released after superplastic deformation at the temperature of 350 ℃ and the strain rate of 5.4×10^{-4}/s, by the elongation of 2500%. Annealing before superplastic deformation releases the stored energy in the specimen, which makes the stress of deformation increase and the elongation decrease.

During superplastic deformation of constant stress, the strain rate is proportion to releasing rate of stored energy. There is an energy parameter, named order of deformation, controlling superplastic deformation, which makes the materials realize uniform deformation. If the change of degree of order brings about the change of strain, then the applied force will increase ordering energy and makes the deformation develop towards increasing the degree of order, which guides the alloy to deform uniformly.

Chapter 3

Microstructures of Zn-Al Alloys

The elongation of Zn-5Al alloy can reach 5000%, and when the composition of Zn-Al alloys diverges from the eutectic point, the elongation of the alloy will abruptly fall, which shows the huge elongation of Zn-5Al alloy is dependent on its eutectic microstructure.

For Zn-22Al eutectoid alloy, after pre-processed for superplasticity, the two-phase structures of equiaxial and fine grains are obtained, and the grain size is 2.5 ± 0.2 μm and the proportion of α to β is 23 against 77 by mass[37].

In this chapter, the microstructures of Zn-Al eutectic alloys in various states are investigated by optical microscope (OM), scanning electron microscope (SEM), transmission electron microscope (TEM), X-ray diffraction (XRD), etc.

3.1 Phase diagram of Zn-Al alloy and its crystallization

Zn-Al phase diagram is generally listed as Al-Zn in the literature. The Al-Zn phase diagram cited in this section is from *Smithells Metals Reference Book*[4].

3.1.1 Phase diagram of Al-Zn alloy

The phase diagram of Al-Zn as shown in Figure 3.1 is often used in the literatures. It is necessary to point out that eutectoid temperature in Figure 3.1 is 275 ℃, while the eutectoid temperature measured by the author using DSC method is 285 ℃. The change is probably due to the addition of 0.04 mass percent Mg in the alloy. Figure 3.2 is a local magnified phase diagram of Al-Zn alloy.

3.1.2 Equilibrium crystallization in Zn-Al alloys

In order to understand the crystallized characteristics of Zn-Al alloys, the equilibrium crystallization for Zn-5Al, Zn-2.5Al and Zn-10Al will be briefly analyzed in this paragraph.

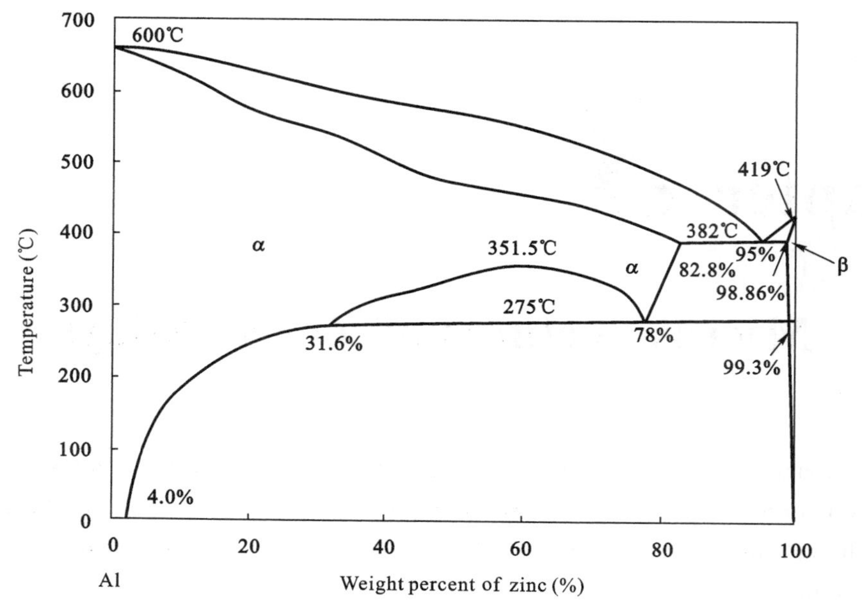

Figure 3.1 Phase diagram of Al-Zn alloy[4].

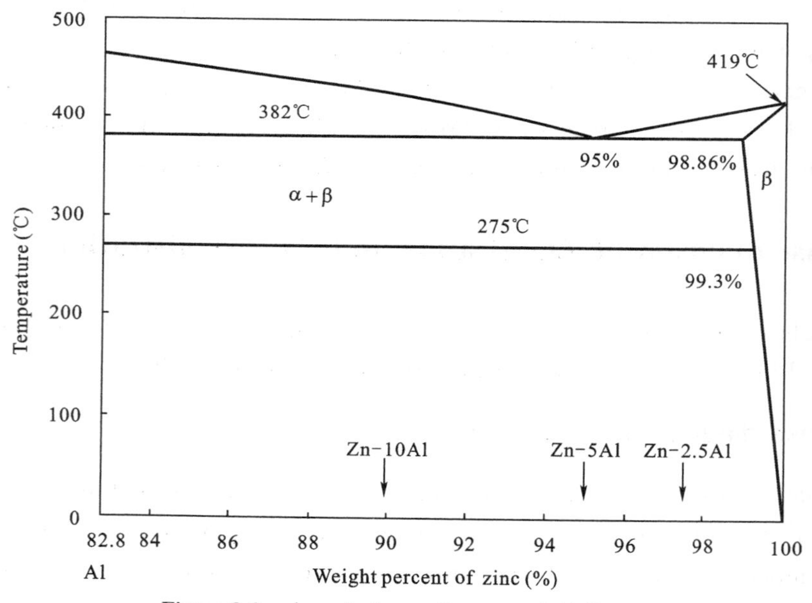

Figure 3.2 Local phase diagram of Al-Zn alloy.

(1) Crystallization in Zn-5Al alloy

For Zn-5Al alloy, an eutectic reaction takes place at 382 ℃, as shown in Figure 3.3. The eutectic reaction may be written as

$$L \rightarrow [24\%\alpha(82.8\%Zn) + 76\%\beta(98.86\%Zn)]_{eutectic}$$

During subsequent cooling, α precipitates from β and β precipitates from α. At the eutectoid temperature, a self-eutectoid reaction takes place, i.e.

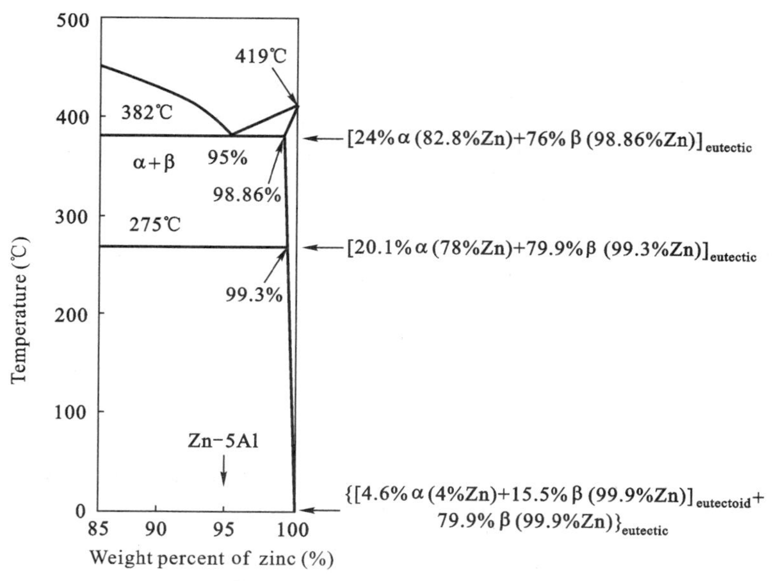

Figure 3.3 Equilibrium crystallization of Zn-5Al alloy.

$$\alpha(78\%Zn) \rightarrow [\alpha(31.6\%Zn)+\beta(99.3\%Zn)]_{eutectoid}$$

Subsequently, α precipitates from β and β precipitates from α again. When cooled to room temperature, the mass percentages of α phase and β phase are 5.2% and 94.8% respectively.

Finally, at the room temperature, the equilibrium structure of Zn-5Al alloy is

$$[(\alpha\beta)_{eutectoid}\beta]_{eutectic}$$

(2) Crystallization of Zn-2.5Al alloy

For Zn-2. 5Al alloy, as shown in Figure 3. 4, before cooling to the eutectic temperature, β_I phase has crystallized from the liquid and its amount reaches 65 mass percent at the eutectic temperature. The remaining liquid alloy will have the same crystallizing processes as in Zn-5Al alloy. At room temperature, the mass percentages of α phase and β phase are 2.5% and 97.5% respectively.

Finally, at room temperature, the equilibrium structure of Zn-2.5Al is

$$\{[(\alpha\beta)_{eutectoid}\beta]_{eutectic}+\beta\}$$

(3) Crystallization in Zn-10Al alloy

For Zn-10Al alloy, as shown in Figure 3. 5, before cooling to the eutectic temperature, α_I phase has crystallized from the liquid alloy and its amount reaches 41 mass percent at the eutectic temperature. The remaining liquid alloy will have the same crystallizing processes as in Zn-5Al alloy. At room temperature, the mass percentages of α phase and β phase are 10.4% and 89.6% respectively.

Finally, the equilibrium structure of Zn-10Al is

$$\{(\alpha\beta)_{eutectoid}+[(\alpha\beta)_{eutectoid}\beta]_{eutectic}+\beta\}$$

It can be concluded that, there are most interface α/β in Zn-5Al alloy, less in Zn-10Al alloy and least in Zn-2.5Al alloy. The tensile results in Section 2.2.1 show that the

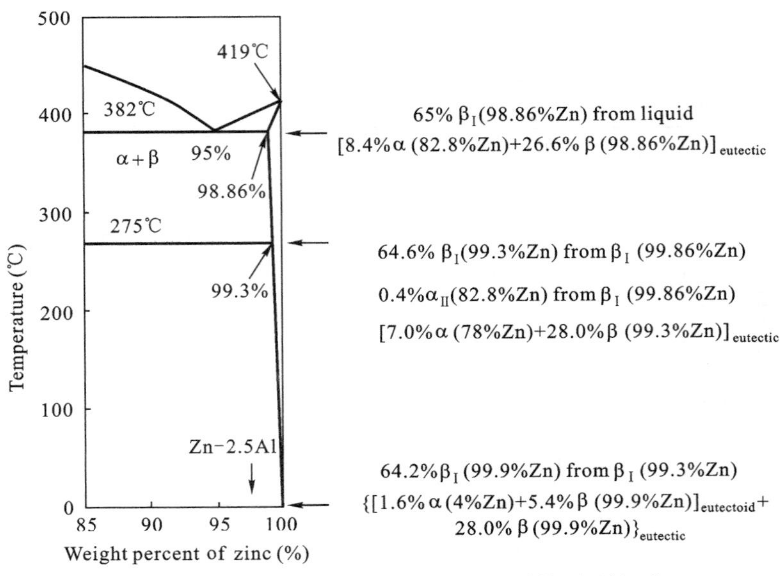

Figure 3.4 Equilibrium crystallization of Zn-2.5Al alloy.

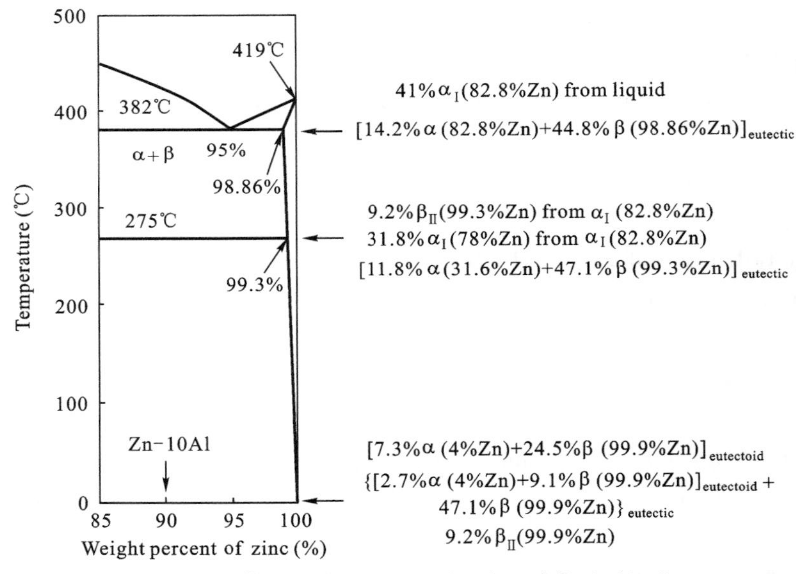

Figure 3.5 Equilibrium crystallization of Zn-10Al alloy.

Zn-2.5Al alloy has the largest deformation stress in three alloys.

3.2 Microstructure of Zn-Al alloys and researching methods

The Zn ingots and the Al ingots with purity of 99.99 mass percent were melted in an intermediate frequency induction furnace. The cast ingots of 300 mm×400 mm×24 mm were obtained by direct chill casting method at 450—500 ℃.

When cooling to 200—300 ℃, the cast ingots were rolled by the deformation of

73%. After annealing by 350 ℃×2 h, the alloy was rolled again, and finally the sheets with thickness of 2 mm or 4 mm were obtained. The compositions of the alloys are shown in Table 3.1 and the impurities in the alloys are shown in Table 3.2.

Table 3.1 Compositions of the alloys (mass percent)

Alloy	Al	Mg	Zn	Al/Zn
Zn-5Al	4.89	0.046	Balance	0.051
Zn-2.5Al	2.52	0.041	Balance	0.026
Zn-10Al	9.17	0.030	Balance	0.100

Table 3.2 Impurities in the alloys (mass percent)

Element	Cu	Pb	Sn	Cd	Fe
Amount	0.0020	0.0027	0.0025	0.0005	0.0099

3.2.1 Investigation by optical microscope

The specimens for metallographic analysis were ground and polished, and then they were etched with an alcohol solution mixed with 5% hydrochloric acid. Figure 3.6 shows the cast microstructures of Zn-5Al, Zn-2.5Al and Zn-10Al alloys. It can be seen from Figure 3.10 that the Zn-2.5Al alloy has the largest grain size in three alloys and that of the proeutectic β is about 100 μm.

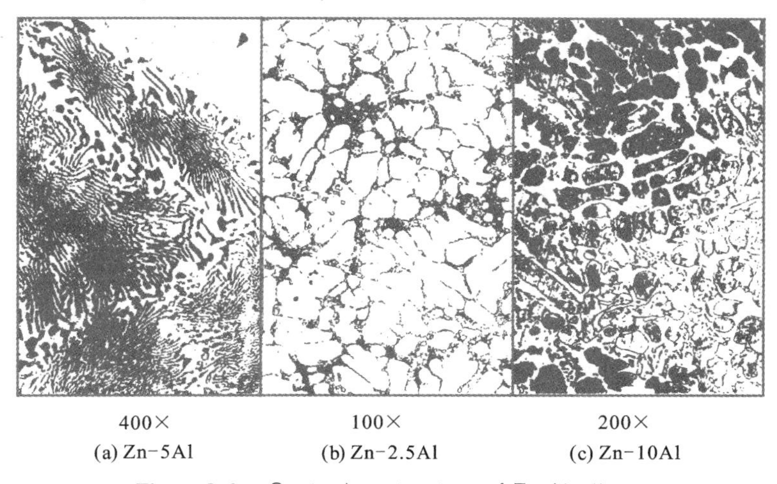

400× (a) Zn-5Al　　100× (b) Zn-2.5Al　　200× (c) Zn-10Al

Figure 3.6 Cast microstructure of Zn-Al alloys.

3.2.2 Investigation by SEM and TEM

In order to improve the contrast of SEM images, a specific etchant was found and it can selectively etch the Zn phase in the Zn-Al alloy, and does not etch the Al phase. The formula of the etchant is: water, 100 mL; HNO_3, 5 mL; Na_2SO_4, 1.5 g; CrO_3, 30 g. With cast specimen of Zn-Al alloys being etched by the etchant, the quasi-three-dimension microstructure with the depth of 10 μm can be seen, as shown in Figure 3.7.

After the Zn-Al alloys are etched, the Zn elements and the Al elements are left in the etchant, and their amounts are analyzed and the result is:

The content of Al in the etchant: 128.0 μg/mL;

The content of Zn in the etchant: 7000.0 μg/mL;

The ratio of content of Zn to Al: 1 : 55.

The specimens for TEM analysis are prepared by electrochemical etching and plasma jet respectively. The specimens are viewed with H-800 transmission electron microscope.

(a) Low amplification (b) High amplification

Figure 3.7 SEM images of cast Zn-5Al alloy.

3.2.3 Other methods

Phases in the alloys were analyzed with Rigaku D/max-RB X-ray diffractometer, and the eutectoid temperature was measured with heat analysis system of Dupont-1090B.

3.3 Experimental results and discussions

3.3.1 Characteristics of Zn-Al alloys in cast state

In cast state, the grain size of Zn-2.5Al alloy is larger than that of Zn-10Al alloy and the microstructure of Zn-5Al alloy is finest in three alloys. Although the compositions of Zn-2.5Al and Zn-10Al alloys diverge slightly from the eutectic point, their microstructures are greatly different from that of Zn-5Al eutectic alloy, as shown in Figure 3.6.

Figure 3.7 is the SEM images of cast Zn-5Al eutectic alloy, and Figure 3.8 is a result obtained by energy dispersive analysis of X-ray (EDAX) under low amplification. It can be seen from the figure that the width and the length of eutectic lamella are about 1 μm and 20 μm respectively.

Figure 3.9 is the TEM images of cast Zn-5Al eutectic alloy. As can be seen from Figure 3.10, the thin is α phase of rich-Al and the wide is β phase of rich-Zn. There is the thinner β between thin α, which comes from the eutectoid reaction at eutectoid temperature and lies in the eutectoid structure. The widths of the α and the β in the eutectoid structure are about 0.08 μm and 0.6 μm respectively.

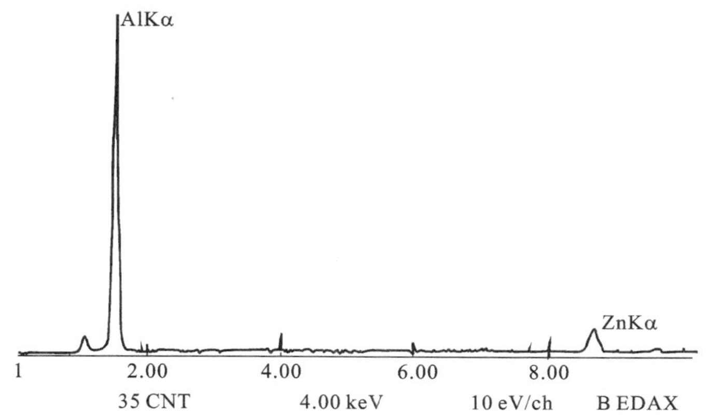

Figure 3.8 EDAX of Zn-5Al alloy etched by etchant under low amplification.

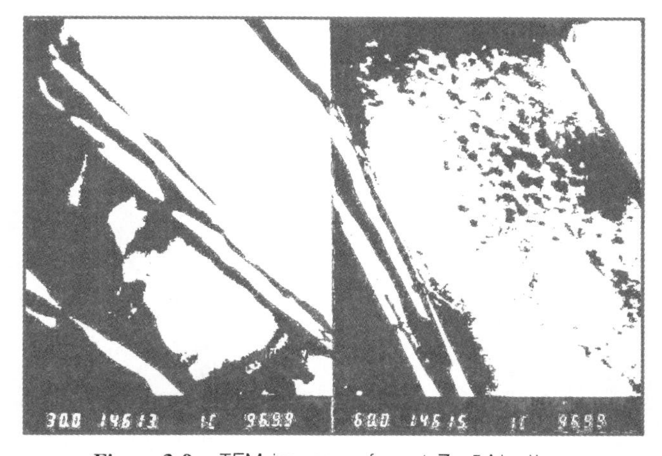

Figure 3.9 TEM images of cast Zn-5Al alloy.

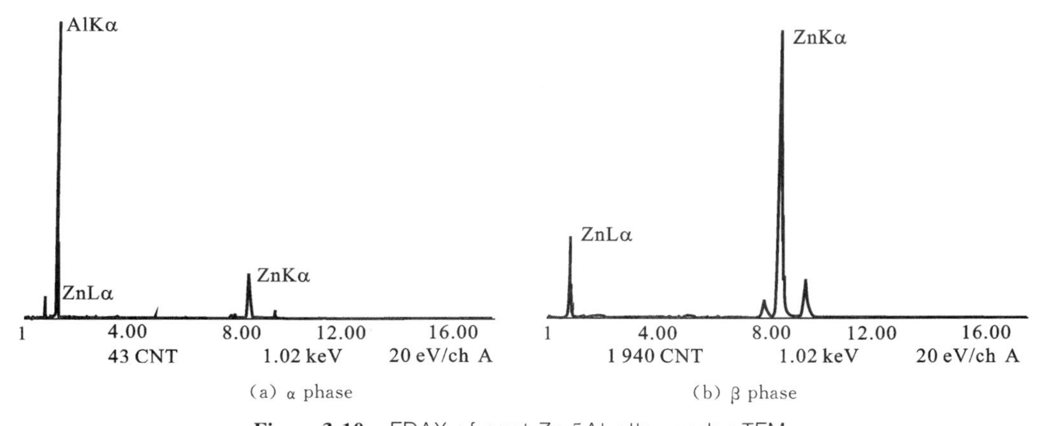

Figure 3.10 EDAX of cast Zn-5Al alloy under TEM.

3.3.2 Characteristics of Zn-Al alloys in rolled and recrystallized states

The specimens for metallographic analysis are cut in the way shown in Figure 3.11, and rolled direction (R), normal direction (N) and transverse direction (T) are marked, and three surfaces corresponding to the three directions are also marked by RS, NS and TS respectively.

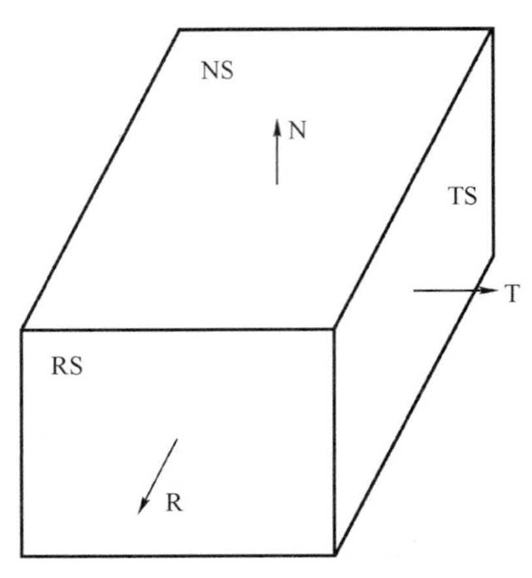

Figure 3.11 Surfaces of the specimen and their representation.

The structure of rolled Zn-5Al alloy exhibits streamlined characteristic along the rolling direction, as shown in Figure 3.12. After short-time recrystallization, the streamline-like microstructure become much more obvious. It can be seen from Figures 3.13 to 3.15 that Zn-5Al alloy has the finest microstructure in three alloys in recrystallized state, and the mean grain size of Zn-5Al, Zn-2.5Al and Zn-10Al alloys is about 3 μm, 12 μm and 6 μm respectively.

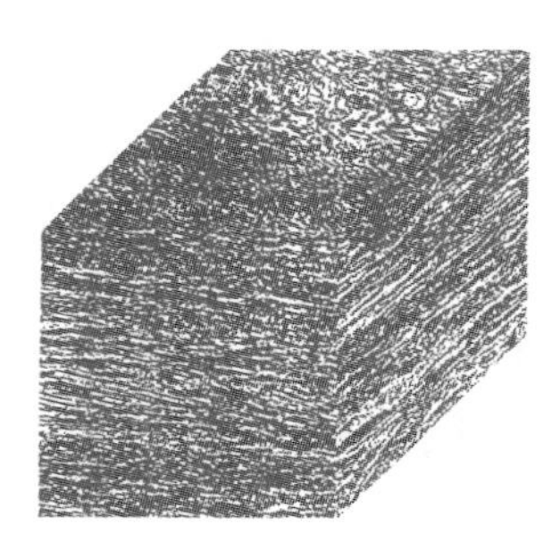

Figure 3.12 The microstructure of rolled Zn-5Al alloy, 428×.

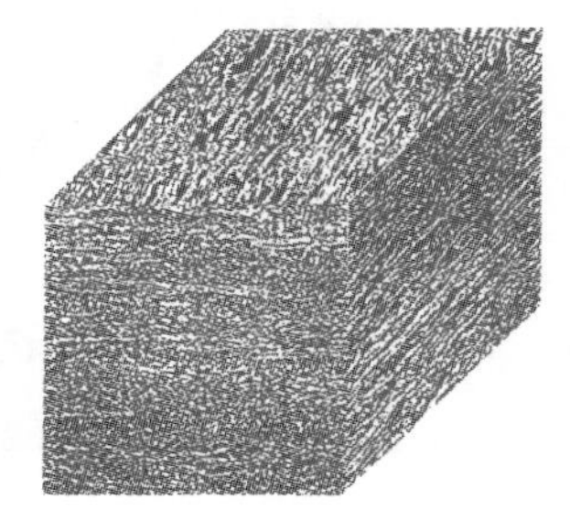

Figure 3.13 The microstructure of Zn-5Al alloy recrystallized by 350 ℃×10 min, 428×.

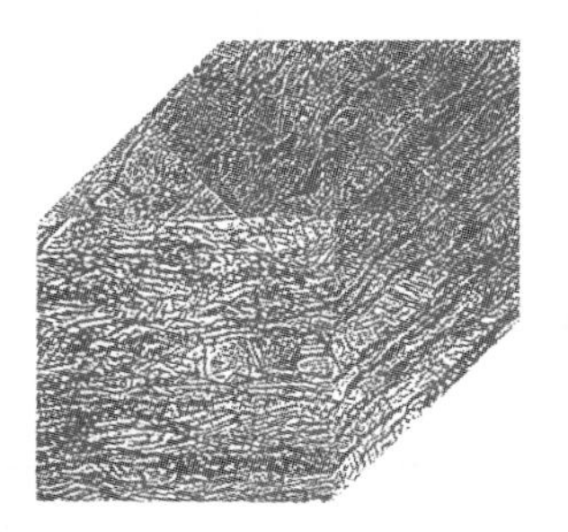

Figure 3.14 The microstructure of Zn-2.5Al alloy recrystallized by 350 ℃×10 min, 428×.

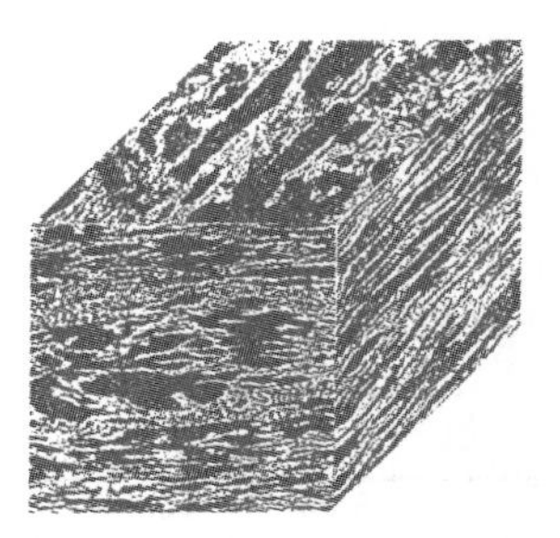

Figure 3.15 The microstructure of Zn-10Al alloy recrystallized by 350 ℃×10 min, 428×.

3.3.3 Phase analysis of Zn-5Al alloy

Table 3.3 is X-ray diffraction data of Zn-5Al alloy recrystallized by 350 ℃×12 h, which shows that there are α phase of rich-Al and β phase of rich-Zn in Zn-5Al alloy, and the peak(111) of α(Al) is about 5% of the peak(101) of β(Zn), which is in accordance with the conclusion in crystallization analysis in Section 3.1.2. In spite of addition of 0.04 mass percent Mg in the alloy, the $MgZn_2$ is too little to be detected by X-ray diffraction.

Table 3.3 X-ray diffraction pattern of Zn-5Al alloy recrystallized by 350 ℃×12 h

No.	2θ	Int.	Width	D	I	Planes
1	36.250	13842	0.450	2.4761	53	Zn(002)
2	38.500	932	0.800	2.3364	4	Al(111)
3	39.000	6262	0.450	2.3076	24	Zn(100)
4	43.200	26285	0.375	2.0924	100	Zn(101)
5	44.750	517	0.450	2.0235	2	Al(200)
6	54.300	4313	0.450	1.6880	16	
7	65.150	282	—	1.4307	1	Al(220)

3.3.4 Microstructure of Zn-5Al alloy deformed superplastically

There is a little change in the microstructure of rolled Zn-5Al alloy during the transformation from recovery to recrystallization, but the elongation and the deformation stress change greatly (see Section 2.2). The recovery prior to superplastic deformation will be discussed in Chapter 4.

The microstructures of Zn-5Al alloy deformed superplastically are shown in Figure 3.16. The grain size grows a little with the increases of elongation and tensile test time. When the elongation is larger than 2000%, many voids appear in the specimens and align along the tensile direction and have a tendency of connect with each other.

3.4 Summary

At between 450 ℃ and 500 ℃, under the condition of iron model and cooling gradually water-cooling, the structure of cast Zn-5Al alloy is fine eutectic lamella, whose breadth is about 1 μm and length is 20 μm. The cast structure of Zn-2.5Al alloy is the coarsest in three alloys, and the proeutectic phase β occurs in the structure by 65% and its grain size is about 100 μm. The cast structure of Zn-10Al alloy is finer than that of the Zn-2.5Al, coarser than that of the Zn-5Al, and the proeutectic phase α occurs in the structure by 41%.

At the eutectoid temperature, the α phase in the eutectic lamellas in Zn-5Al alloy has an eutectoid reaction, and the eutectoid lamellas formed in the reaction have α phase

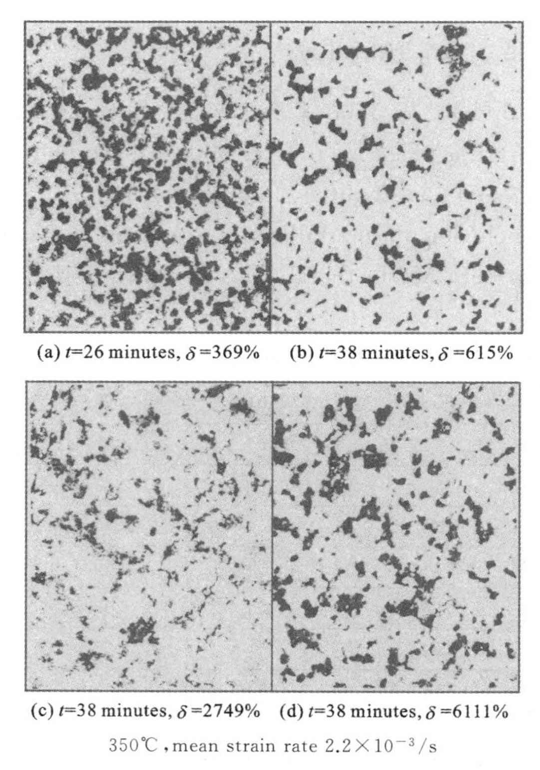

(a) t=26 minutes, δ=369%　(b) t=38 minutes, δ=615%

(c) t=38 minutes, δ=2749%　(d) t=38 minutes, δ=6111%

350℃, mean strain rate 2.2×10^{-3}/s

Figure 3.16 Microstructure of Zn-5Al alloy deformed superplastically, 400×.

of 0.08 μm and β phase of 0.6 μm in width. The eutectoid lamellas are parallel to the eutectic lamellas and both have the same length.

There are α/β-type interfaces in Zn-5Al, Zn-2.5Al and Zn-10Al alloys by 100%, 35% and 59% respectively. The elongation of Zn-5Al alloy can reach 5000% in superplastic deformation, which is dependant on α/β-type interfaces in the alloy.

Chapter 4

Grain Boundary Sliding Controlled by Diffusion Dissolution Zone

The elongation of the alloy can reach 5000%, when Zn-5Al is deformed superplastically at 350 ℃, which is related with α/β-type interface in the alloy and its behavior of sliding. In this chapter, firstly, the grain boundary sliding model of superplasticity is introduced. Then, the experimental methods and results observing grain boundary sliding are stated. Finally, for Zn-5Al eutectic alloy, the mechanism of phase boundary sliding controlled by diffusion dissolution zone will be discussed.

4.1 Grain boundary sliding model of superplasticity

Since superplasticity was discovered, hundreds of superplastic materials have been developed. Several models concerning the mechanism of deformation have been proposed for different materials, such as dissolution-precipitate theory[38], metastable state theory[39], diffusion flow mechanism[40], dislocation creep theory[41], grain exchange model[42] and grain rearrangement[43]. Generally, equiaxial and fine grains are considered to be an essential condition for realizing structural superplasticity. At present, theories about superplastic deformation have focused on grain boundary sliding (GBS) models, among which two models are worthy to be introduced, viz., Ball-Hutchson model accommodated by dislocation and Ashby-Verall model accommodated by diffusion.

4.1.1 GBS model with dislocation accommodation

Grain boundary sliding model with dislocation accommodation is also called Ball-Hutchson model[44]. Ball and Hutchson observed the microstructure of Zn-22Al alloy deformed superplastically with TEM and found pile-up of dislocations in α phase. They thought that the slip accommodation process involves the sequential steps of glide and climb. When climb is the rate-controlling step, there is the pile-up stress at the head of the climbing dislocation. When glide is the rate-controlling step, there is no pile-up stress. As shown in Figure 4.1, stress concentration makes it possible for dislocation in the grain to slide and pile up at the grain boundary. Under the action of stress concentration, dislocation at the head of pile-up of dislocations climbs into the grain boundary with the movement along the grain boundary until annihilation. During this process, due to the accommodation of grain rotation, there is no obvious grain extension.

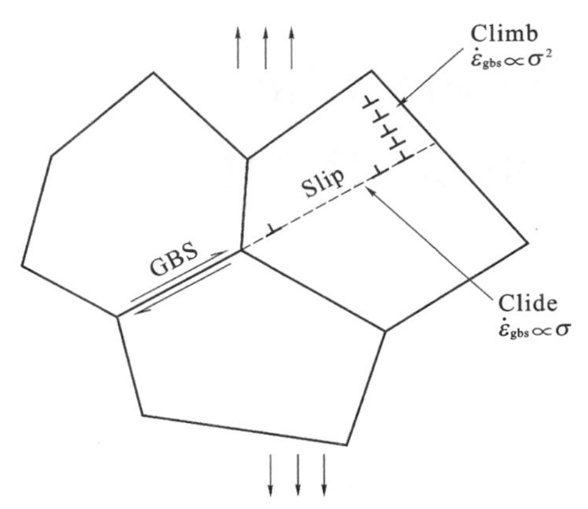

Figure 4.1 Ball-Hutchson model with dislocation accommodation[44].

4.1.2 GBS model with diffusion accommodation

Grain boundary sliding model with diffusion accommodation is also called Ashby-Verall model[45]. In this model, it is believed that large strain is the result of grain exchange and a two-dimension grain exchange pattern is proposed, as shown in Figure 4.2. Ashby and Verall thought that work done by applied force is dissipated in four irreversible processes: diffusion-accommodation flow; the increasing of grain boundary areas; driving grain boundary as the source and well of cavitation; grain boundary sliding overcoming viscosity. Apparently, Ashby-Verall model is a two-dimension model, and it cannot explain the increasing of new surface.

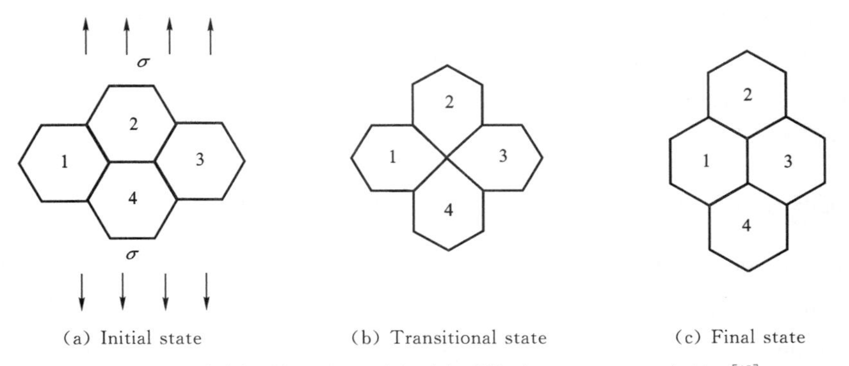

Figure 4.2 Ashby-Verall model with diffusion accommodation[45].

4.2 Experimental methods of viewing GBS

First, the surface of specimens is polished, and some specimens are electroplated with Zn, Cu and Ni respectively. Then, the specimens with electroplating and without electroplating are deformed superplastically. Finally, the grain boundary sliding in the two types of specimens are viewed.

4.2.1 Investigation of surface of polished specimen

Before superplastic deformation, the surface of specimens of Zn-2.5Al alloy was polished with Al_2O_3 with a grain size of 1 μm. The superficial appearance of the deformed specimens is shown in Figure 4.3. The figure (a) shows the surface of the specimen deformed in an electric field for 50 hours with an elongation of 12%. Grain boundary occurs on the surface of specimen, which can be clearly observed under SEM. The figure (b) shows the surface of the specimen deformed without electric field for 50 hours with an elongation of 92%. As can be seen, very obvious sliding characteristics occur on the surface of the specimen. Suppression of electric field to phase boundary sliding will be discussed in Chapter 6. It is shown by energy dispersive analysis of X-ray (EDAX) that non-deformed zone ① contains mainly the Zn, as shown in Figure 4.4(a), and deformed zone ② contains both the Zn and Al, as shown in Figure 4.4(b).

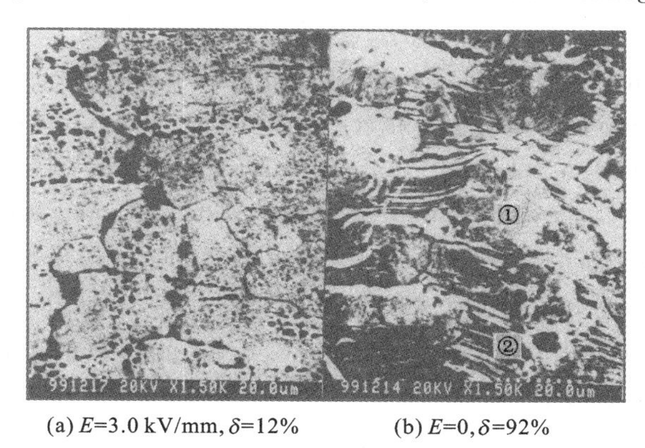

(a) E=3.0 kV/mm, δ=12%　　(b) E=0, δ=92%

275 ℃, initial stress 2.23 MPa, time of deformation 50 hours

Figure 4.3 Superficial appearance of Zn-2.5Al alloy after superplastic deformation.

The microstructure of Zn-2.5Al alloy is $(\alpha+\beta)+\beta$, where β is coarse proeutectic phase and $(\alpha+\beta)$ is finer eutectic. Deformed zone with Zn and Al corresponds to eutectic $(\alpha+\beta)$ and non-deformed zone with only Zn corresponds to β phase. It is indicated by the experimental results that $(\alpha+\beta)$ in Zn-2.5Al alloy is easy to deform and proeutectic β is not easy to deform.

For Zn-5Al alloy, due to its microstructure of entire eutectic $(\alpha+\beta)$, the deformation stress is relatively low and the grain boundary sliding is comparatively homogeneous, as shown in Figure 4.5.

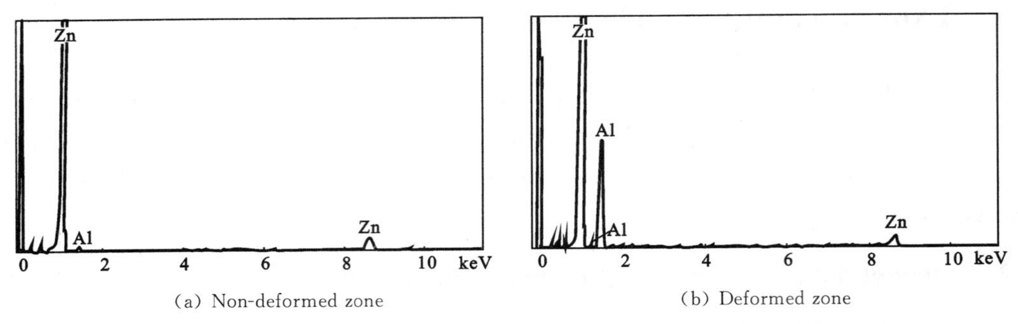

(a) Non-deformed zone (b) Deformed zone

Figure 4.4 EDAX of surface of Zn-2.5Al alloy after superplastic deformation.

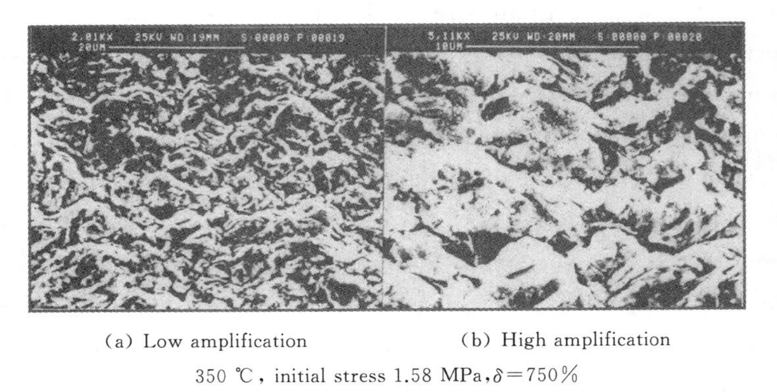

(a) Low amplification (b) High amplification

350 ℃, initial stress 1.58 MPa, $\delta=750\%$

Figure 4.5 Superficial appearance of Zn-5Al alloy after superplastic deformation.

Ball-Hutchson model is a short-distance sliding model of grain boundary. Grain boundary sliding hampered by grains needs to be accommodated by the glide of dislocations for succession. The deformation of Zn-2.5Al alloy is just this case.

For Zn-5Al alloy, the grain boundary sliding during superplastic deformation is a long-distance sliding, as shown in Figure 4.6: A and B represent two long-distance paths of grain boundary sliding. First, the deformation occurs along path A, as shown in the figure (b). After cessation of deformation through path A, deformation takes place along path B, as shown in figure (c). As can be seen, there is no need to accommodate by dislocations during the superplastic deformation. Results of sliding are that the specimen is lengthened in the tensile direction and contracted in the direction perpendicular to the tensile axis. In the mean time, the specimen increases new surfaces. When the elongation reaches 5000%, the external surfaces will increase by 10 times. If the surface of the specimen is electroplated, the sliding on the surface will be resisted by the electroplated layer, which will influence deformation of the specimen. For Zn-5Al alloy, the long-distance sliding of grain boundary is controlled by a diffusion dissolution zone, which will be discussed in the next section.

4.2.2 Investigation of surface of electroplated specimen

The specimens of Zn-5Al alloy in rolled state are electroplated with Zn, Cu and Ni

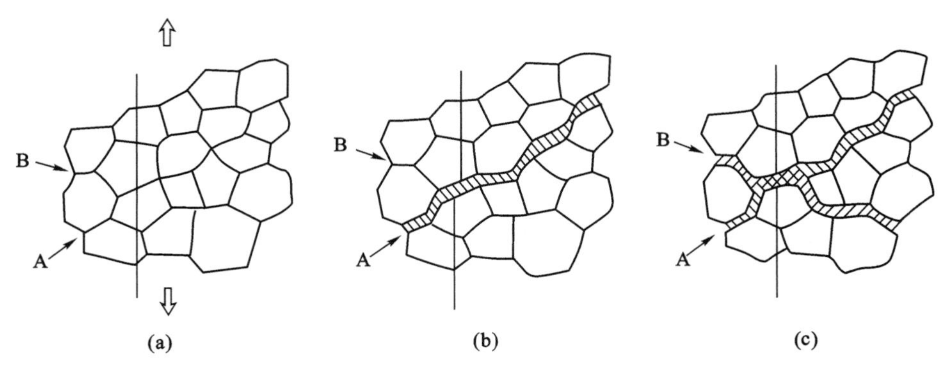

Figure 4.6 Grain boundary long-distance sliding for superplastic deformation.

respectively. The electroplating layer on the specimens has the thickness of 5 μm to 10 μm, and its structure is different from that of the matrix. It is shown by X-ray diffraction that the layers electroplated with Zn, Cu respectively are completely in crystalline state. Figure 4.7 is the X-ray diffraction pattern of electroplated-Cu specimen. The electroplating layer has certain mechanical resistance to superplastic deformation in Zn-5Al alloy. Figure 4.8 is the superplastic tensile results of electroplated specimen. When the elongation of deformation reaches 60%, the strain rate of non-electroplated specimen is 4.2×10^{-3}/s, and the strain rates of the specimens electroplated with Zn, Cu and Ni are 8.3×10^{-4}/s, 4.2×10^{-4}/s, 2.1×10^{-3}/s respectively. As can be seen, the strain rate of superplastic deformation is decreased by the electroplating layer on the surface of specimens.

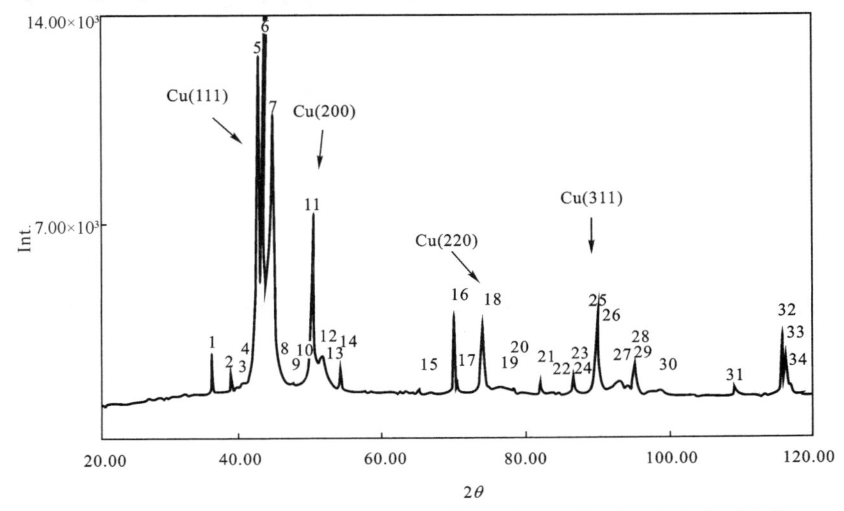

Figure 4.7 X-ray diffraction pattern of specimen electroplated with Cu.

Figure 4.9 is the superficial appearance of electroplated-Cu specimen. The figure (a) shows that the specimen is lengthened along the tensile axis and contracted in the direction perpendicular to the tensile axis. The figure (b) shows the patterns of deformation of the matrix under the electroplating layer. It can be seen that the flowing

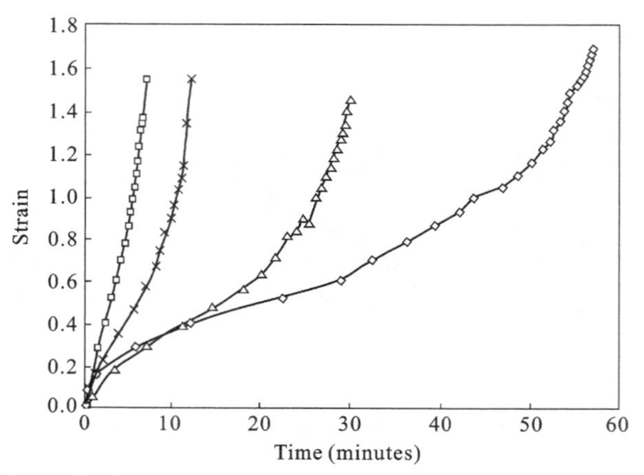

□ Non-electroplated, × Electroplated Ni, △ Electroplated Cu, ◇ Electroplated Zn

350 ℃, initial stress 0.8 MPa

Figure 4.8 Effect of electroplating layer on superplasticity.

matrix has a rise and fall as wave. The figure (c) shows the effect of the broken electroplating layers on the flow of the matrix. The long streamline shows the sliding to be long-distance.

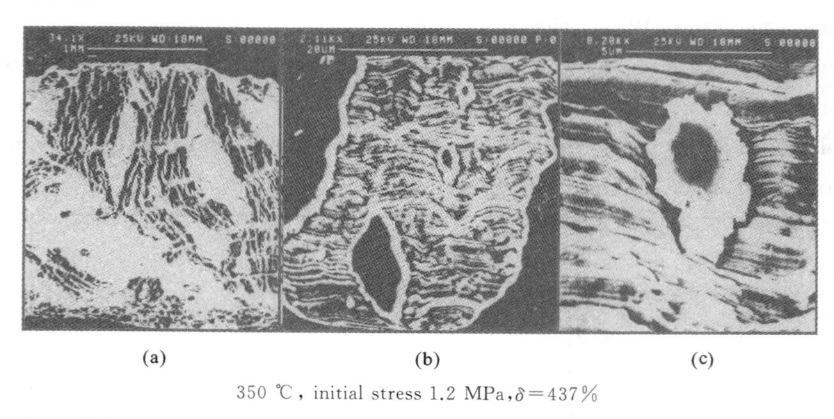

350 ℃, initial stress 1.2 MPa, $\delta = 437\%$

Figure 4.9 Superficial appearance of superplastic deformation of specimen electroplated with Cu.

4.3 Phase boundary sliding controlled by diffusion dissolution zone

All grain boundaries in Zn-5Al alloy are almost phase boundaries, i. e., α/β interfaces. Obviously, α phase and β phase have more α/β interfaces in lamella shape than in spherical shape. Rolled deformation makes α phase and β phase in some orientation break, but the amount of α/β interfaces would not be decreased.

4.3.1 Diffusion and dissolution in Al/Zn interface

A eutectoid microstructure in Zn-Al system can be obtained by reaction method of Zn powder and Al powder. First, the Al powder of 5 mass percent and the Zn powder of 95 mass percent are mixed and the mixture is shaped at 300 ℃ under the pressure of

1700 MPa. Then, the powder-made specimens are sintered at 350 ℃ for long time. Figure 4.10 shows eutectoid microstructure obtained by sintering method. The figure (a) shows the shape of pure Al powder. The appearances of the specimens sintered for 62, 80 and 100 hours are shown in the figures (b), (c) and (d) respectively. It is found by experiments that the eutectoid microstructure always occurs in Al grains and there is no the eutectoid microstructure in Zn grains. When the powder-made specimens are kept at 350 ℃ for 62 hours, Zn atoms dissolve in Al grains through diffusing, which results in a solid solution of Zn in Al, α'. In subsequent cooling process, there is an eutectoid reaction in α' at the eutectoid temperature of Zn-Al system, which produces the eutectoid microstructure ($\alpha+\beta$). The figures (c) and (d) are the results of diffusion treatment for 80 and 100 hours. As can be seen, the extent of the eutectoid microstructure is limited by the grain size of Al powder.

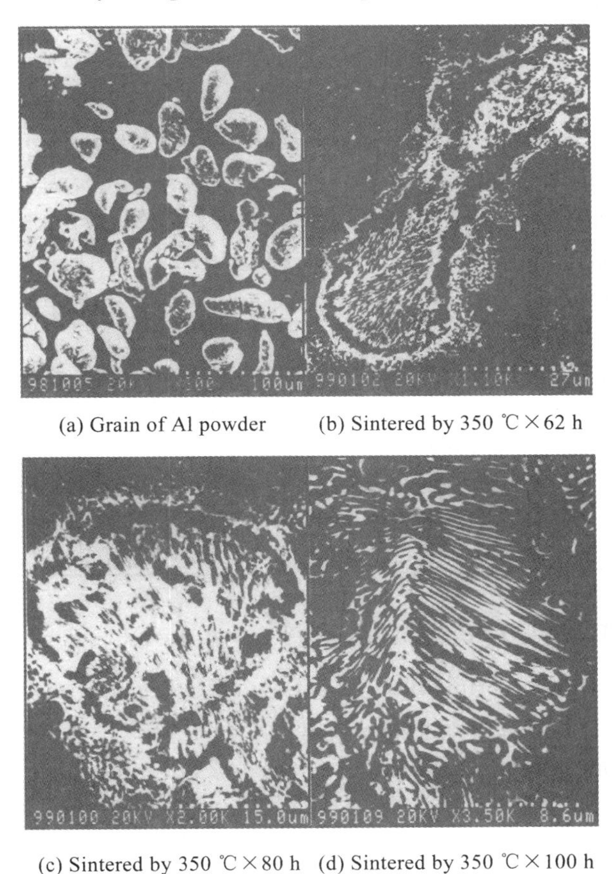

(a) Grain of Al powder (b) Sintered by 350 ℃×62 h

(c) Sintered by 350 ℃×80 h (d) Sintered by 350 ℃×100 h

Figure 4.10 Eutectoid microstructure of powder-made specimen.

In fact, it is an application of Kirkendall effect[46] that the eutectoid microstructure is obtained in powder-made specimen of Al powder and Zn powder through atomic diffusion at high temperature. Kirkendall effect of Ni/Cu diffusion couples is introduced in the literature [46]. In spite of unlimited solid solution formed by Ni and Cu, in the case of solid state diffusion, Cu can diffuse into Ni, but Ni cannot diffuse into Cu. Table 4.1 is the comparison of some parameters of Kirkendall effect between Al/Zn and Ni/Cu.

Table 4.1 The comparison of Kirkendall effect between Ni/Cu and Al/Zn

	Ni/Cu	Al/Zn
Coefficient of diffusion	$D_{Ni}<D_{Cu}$	$D_{Al}<D_{Zn}$
Bonding energy	$E_{Ni}>E_{Cu}$	$E_{Al}>E_{Zn}$
Direction of diffusion	Cu into Ni	Zn into Al
Atomic radius	$r_{Ni}<r_{Cu}$	$r_{Al}>r_{Zn}$
Electron density	$n_{Ni}>n_{Cu}$	$n_{Al}<n_{Zn}$

4.3.2 Diffusion dissolution zone in Al/Zn interface

The principle regarding the formation of the eutectoid structure from Al powder and Zn powder is illustrated in Figure 4.11. During the sintering process, a diffusion dissolution zone is formed in Al/Zn interface, as shown in the figure (a). The diffusion dissolution zone is nonhomogeneous solid solution α', which comes from the dissolution of Zn into Al. α' has same crystal structure as α in Zn-5Al, while the contents of Zn in α' is larger than that in α at room temperature. In addition, the distribution of Zn in α' is not uniform, for there is little Zn at the Al end and 65 atomic percent at the Zn end. In subsequent cooling process, α' transforms into eutectoid structure ($\alpha+\beta$) at the eutectoid temperature, as shown in the figure (b).

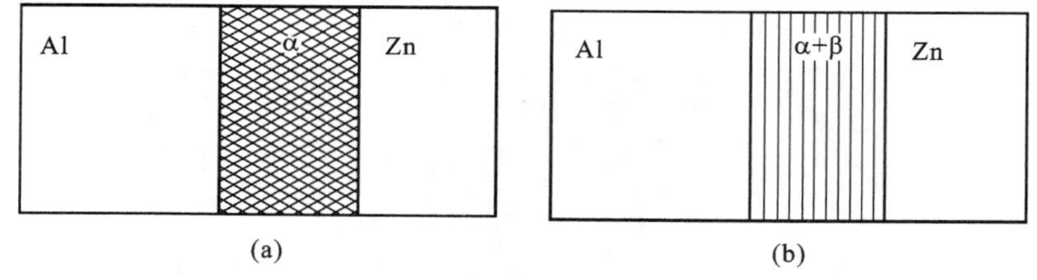

Figure 4.11 Diffusion dissolution zone in Al/Zn interface.

From the sintering experiments, the following inferences can be obtained:

(1) Above the eutectoid temperature, Kirkendall effect will occur in Zn-5Al alloy, i.e., α phase becomes wider and β phase becomes thinner. Because the α phase is originally thinner and β phase is originally wider, the average size of grains remains unchanged. This is just the cause that Zn-5Al alloy has a stable grain size at superplastic deformation temperature.

(2) At 350 ℃, the crystal structure of Al is more stable than that of Zn. Zn can dissolve into Al and Al cannot dissolve into Zn, which shows that Zn atoms contacting Al atoms in interface Zn/Al have weaker bonding force with the original Zn crystal.

(3) At 350 ℃, there are some Zn atoms easy to move in interface α/β. These atoms have weaker bonding force with β, and so is the case with α before dissolution into α.

Consequently, if certain external force acts on the specimen along interface α/β, interface α/β is very easy to slide.

(4) At 350 ℃, it will take 50 hours to get saturated solid solution α' with grain size of 5 μm. Hence the migration speed of phase boundary is 0.0017 μm/min, according to an equation, $v = l/(2t)$, where l is the migration length of phase boundary, t is the migration time[46]. The result of migration is that interface Al/Zn moves toward to Zn end.

4.3.3 Mechanism of phase boundary sliding controlled by DDZ

(1) Diffusion dissolution zone advantageous to sliding

Figure 4.12 is a tensile test result of Zn-5Al alloy, which is obtained by means of relaxation. For a commercial materials test machine, the displacement of load cell on it is generally very small. Therefore, the tensile test using such a machine is rigid extension, i. e., the lengthening of specimen is the main contribution to the displacement of grips.

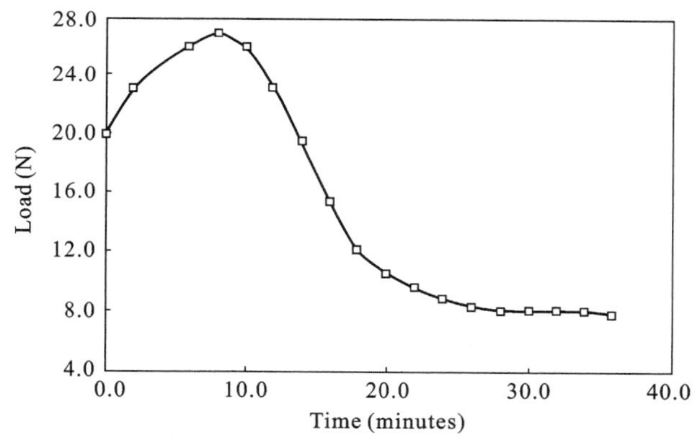

Figure 4.12 Result of relaxed tensile test of Zn-5Al alloy.

However, the load cell used here is just a spring with a small force constant of 0.26 N/mm. The lengthening and pulling force of the spring are shown in Table 4.2. At the beginning of the test, the lengthening of spring is the main contribution to the displacement of grips; hence it is a relaxed tensile test.

Table 4.2 Lengthening and pulling force of the spring

Lengthening (mm)	20	40	60	80	100
Pulling force (N)	2.6	10.4	15.6	20.8	26.0

When test temperature reaches 350 ℃, the tensile test begins with a pre-tension of 20 N(0.6 MPa). In 8 minutes, the pulling force reaches its maximum of 27 N(1.8 MPa). The spring is lengthened by 25 mm and the specimen is lengthened by 100%. As discussed in Section 4.3.2, the diffusion dissolution zone in the specimen is less than 0.02 μm in thickness. Decreasing of pulling force is reduction of lengthening of the spring, which means that on the one hand, one end of the specimen moves at a rate of the machine's grip, and on the other hand, the other end of the specimen moves in the opposite direction along with the spring. At this time, the stress decreases and strain rate significantly increases, as shown in Figure 4.13. The increasing of strain rate in the figure corresponds to the sliding of diffusion dissolution zone on microscope scale.

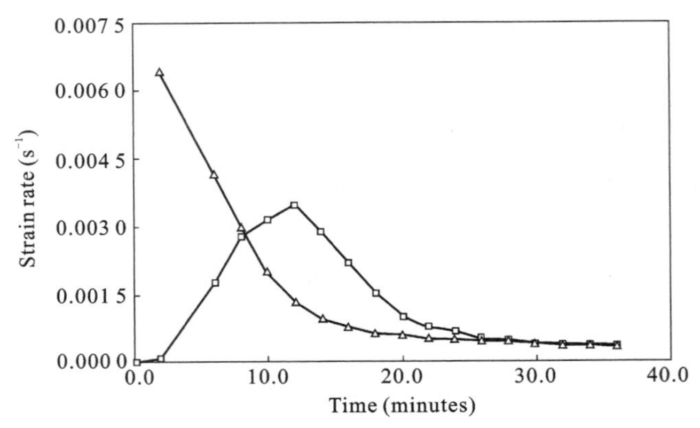

△ Contribution of grip to strain rate, □ Total strain rate of specimen

Figure 4.13 Relaxed deformation and sliding of diffusion dissolution zone.

For rolled Zn-5Al alloy, high defect concentration is advantageous to the diffusion of atoms. Otherwise, diffusion dissolution zone, namely, α' solid solution with more defects, is easy to occur in interface α/β, whose concentration of solute Zn is about 1 atomic percent at the α end and is x atomic percent at the β end, where $x<65$, viz., the dissolubility of Zn in Al at 350 ℃. This kind of unsaturated diffusion dissolution zone with more defects has weaker bonding force with β phase and is advantageous to sliding, as shown in Figure 4.14(b).

(2) Diffusion dissolution zone disadvantageous to sliding

Annealing makes superplasticity in Zn-5Al alloy decrease. Zn-5Al alloy in rolled state has the largest elongation in superplastic tensile test. After temporary recrystallization, for example, at 350 ℃ for 10 — 20 minutes, the stress abruptly increases and the elongation suddenly decreases. As discussed in Section 3. 3, the microstructure of the specimen after recrystallization treatment at 350 ℃ for 10 minutes has little difference from that in rolled state, consequently, the heat treatment by 350 ℃ × 10 min is similar to recovery. During the course of recovery, stored energy in the alloy is released greatly.

The specimens are annealed at the temperature of superplastic deformation prior to tensile test, which makes diffusion dissolution zone thicken. With the annealing of 20 minutes, diffusion dissolution zone increases a little, as shown in Figure 4.14(c); with the annealing of 2.5 hours, diffusion dissolution zone increases more. Assume that the dissolution of Zn in α' reaches dynamic saturation, i.e., the solubility of Zn in α' near α end is about 1 atomic percent and that near β end is 65 atomic percent, which is solubility of Zn in Al at 350 ℃.

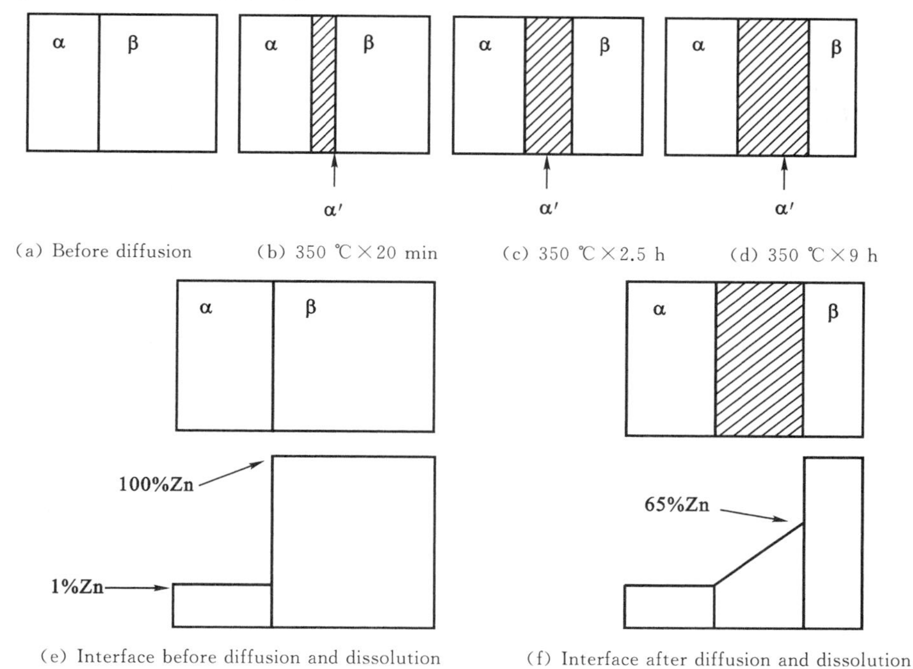

(a) Before diffusion (b) 350 ℃×20 min (c) 350 ℃×2.5 h (d) 350 ℃×9 h

(e) Interface before diffusion and dissolution (f) Interface after diffusion and dissolution

Figure 4.14 Schematic illustration of diffusion dissolution zone.

The time of α' reaching dynamic saturation is called critical time, as shown in Figure 4.14(f). With annealing of 9 hours, the thickness of diffusion dissolution zone continuously increases. When β precipitates from α', the following orientation relation is satisfied[47-48]:

$$(111)_{\alpha'} // (001)_{\beta} \tag{4-1}$$

α' and α have the same face-centered cubic (FCC) crystal structure, and β has hexagonal close-packed (HCP) structure. The crystal lattice of α' is the epitaxy of the lattice of α, and their difference only lie in that α' has more Zn than α. Hence interface α/α' is disadvantageous to sliding. Assume that interface α'/β satisfies Equation (4-1), i.e., the plane (001) in β phase is the epitaxy of the plane (111) in α' phase. For α' phase of FCC structure, the stacking sequence of plane (111) is ABCABC, and for β phase of HCP structure, the stacking sequence of plane (001) is ABABAB. At β end of α', Zn in α' reaches 65 atomic percent. Under this special condition, it seems that it is Zn atoms that construct FCC structure of α' at 350 ℃. Obviously, the transition from FCC structure to HCP structure is just that from ABCABC to ABABAB. Consequently, the saturated interface α'/β is disadvantageous to sliding.

Annealing treatment on Zn-5Al alloy prior to superplastic deformation increases the thickness of diffusion dissolution zone in interface α/β, which is disadvantageous to the sliding of α/β. When the Zn of diffusion dissolution zone reaches dynamic saturation, diffusion dissolution zone has the largest effect on superplasticity and further prolonged annealing time would not have significantly increasing effect on stress and elongation of superplastic deformation.

4.4 Summary

The structure of Zn-2.5Al alloy is $(\alpha+\beta)+\beta$. During superplastic deformation, eutectic $(\alpha+\beta)$ mainly containing interface α/β is advantageous to sliding and proeutectic β phase containing interface β/β is disadvantageous to sliding.

Interface α/β in Zn-10Al alloy is less than that in Zn-5Al alloy and the superplasticity of Zn-10Al alloy is inferior to that of Zn-5Al alloy. Interface α/β in Zn-5Al alloy can realize long-distance sliding, i.e., sliding can penetrate onto the surface of the specimen. Electroplating layer on the surface can influence phase boundary sliding and finally influence superplastic deformation of the whole specimen.

At superplastic deformation temperature, higher defect concentration in rolled Zn-5Al alloy makes atoms easily migrate, which promotes diffusion dissolution zone to form in interface α/β. The unsaturated diffusion dissolution zone, viz., the nonhomogeneous solid solution α' with the solute of Zn near β less than 65 atomic percent, is advantageous to sliding of interface α/β.

With an increase of annealing time before superplastic deformation, diffusion dissolution zone in interface α/β will reach saturation with the solute of Zn being 1 atomic percent at α end and 65 atomic percent at β end. The crystal structure of diffusion dissolution zone at α end is the epitaxy of α phase, hence diffusion dissolution zone has stronger bonding force with α and is disadvantageous to sliding; in spite of FCC structure of diffusion dissolution zone at β end, it is composed of 65 atomic percent Zn atoms, if the process of β dissolving into α' satisfies the same orientation relation as the process of β precipitating from α', then transition of crystal structure from diffusion dissolution zone to β phase is just the transition of stacking sequence from ABCABC to ABABAB, therefore the interface between saturated diffusion dissolution zone and β is also disadvantageous to sliding.

Before superplastic deformation, the effect of annealing time on elongation and stress has a critical point, and the annealing exceeding the critical time no longer affects superplasticity remarkably. This is corresponding to the state that the distribution of Zn in diffusion dissolution zone reaches dynamic saturation, and continuous annealing only increases the thickness of diffusion dissolution zone and its bonding with two-side phases remains unchanged.

Chapter 5

Evolution of Diffusion Dissolution Zone

Zn-5Al alloy in rolled state has certain stored energy and an unsaturated diffusion dissolution zone is easy to form in its interface α/β, which is advantageous to superplastic deformation. With the development of superplastic deformation, stored energy in the alloy becomes less and less, and the lattice distortion and the lattice disarrangement produced by cold work gradually become eliminated and the lattice constant also changes to some extent.

In this chapter, first two kinds of electron theories of alloy are introduced. Then X-ray diffraction method for alloy phase is remarked. After that the changes of the lattice constants of Al phase and Zn phase in Zn-5Al alloy are analyzed. Finally the experimental results are explained according to the electron theory of alloy and the contribution of the strain of lattice cell to superplastic deformation is discussed.

5.1 Electron theories of alloy

In this section, first empirical electron theory founded by Yu Ruihuang is introduced. Then, improved TFD model proposed by Cheng Kaijia is narrated.

Electron structure of metallic materials is always the object of researches in the field of quantum mechanics and solid physics. It is in the Hume-Rothery's empirical law of the concentration of the valence electron that concept of the electron structure was first used in the study of alloys[49]. Chemical bond theory developed by Pauling reveals the quantum mechanics background of valence bond characteristics, which prompts the development of electron theories of alloy.

5.1.1 Yu's empirical electron theory

In 1978, Yu Ruihuang founded "Empirical Electron Theory of Solid and Molecules (EET)"[1-2]. In this theory, first construct two atomic states or so-called h state and t state, then get hybridized states of atoms according to the corresponding formula and distribute the electrons among the bonds, and finally calculate bond length using revised

Pauling formula and obtain so-called theoretical bond length. On the other hand, calculate some distances between atoms according to the lattice constant and obtain so-called experimental bond length. At last make a comparison between the theoretical bond length and the experimental bond length. If the difference is less than 0. 05 angstrom, believe that constructed atomic states, i. e., electron structures are reasonable, otherwise, carry out the above calculation again until theoretical bond length and experimental bond length are in good agreement.

The essence of EET is the amount of bonding electrons. All the bond length can be calculated according to crystallographic data. For example, in the Al crystal of FCC structure, there are 12, 6 and 24 bonds in the direction <110>, <100> and <112> respectively. The amount of the valence electron of the crystal of Al in the <110> and <100> directions given by EET is listed in Table 5.1.

Table 5.1 The valence electron structure in Al crystal

Hybridized state	4	5
Electron number in <110>	0.20857	0.23886
Electron number in <100>	0.00446	0.00510

According to EET, there are 6 hybridized electron structures of Al crystal and two of them satisfy the 0. 05 criterion. There are 18 hybridized electron structures of Zn crystal and nine of them satisfy the 0.05 criterion, as shown in Table 5.2. In practical calculation, the choice of hybridized states is a problem that cannot be answered completely by EET. This is the so-called multi-solution problem of EET.

Table 5.2 The valence electron structure in Zn crystal

State	Dist.1	Dist.2	Dist.3	Dist.4	Dist.5
10	0.40783	0.15595	0.00296	0.00023	0.00006
11	0.39522	0.15113	0.00287	0.00022	0.00006
12	0.39138	0.14966	0.00284	0.00022	0.00006
13	0.37989	0.14526	0.00276	0.00021	0.00006
14	0.37540	0.14355	0.00273	0.00021	0.00006
15	0.36907	0.14113	0.00268	0.00021	0.00006
16	0.36489	0.13953	0.00265	0.00020	0.00006
17	0.36190	0.13839	0.00263	0.00020	0.00006
18	0.35861	0.13856	0.00261	0.00020	0.00006

Note: dist.1 is the shortest bond, dist.1<dist.2<dist.3<dist.4<dist.5.

5.1.2 Improved TFD model

In 1991, Cheng Kaijia proposed an improved TFD model (TFDC) for materials design[50]. The electron density of the atom's surface in TFDC model is given when the atoms form metal. In fact, Wigner-Seitz atomic radius is adopted in TFDC model and the electron density is the number of electrons per unit volume at the radius. For example, the atomic radius and electron density of Al and Zn are 1.58×10^{-8}cm, 1.55×10^{-8}cm and $1.05\times10^{23}/cm^3$, $1.86\times10^{23}/cm^3$ respectively. The Wigner-Seitz atomic radius and TFDC electron density at this radius for some pure metals are listed in Table 5.3. In TFDC model, the bonding electrons are evenly distributed on the atom's surface without considering the orientation of bonds.

TFDC model has an important boundary condition: electron density between atoms in the interface should be continuous, which ensures the continuity of wave function required by quantum mechanics[50]. Consequently, when atoms of different kinds form solid, the radii of atoms are bound to change to satisfy the equality of electron density between atoms in the interface. For crystal, the change of atomic radius means the change of lattice constant. The boundary condition reveals that the interaction of electrons with lattice is an important phenomenon universally existing in the crystalline materials.

Table 5.3 Atomic radius and electron density on atom's surface of common metals

Element	Ni	Co	Cu	Au	Zn	Al
Atomic radius(10^{-10}m)	1.38	1.39	1.41	1.59	1.55	1.58
Electron density($10^{29}\cdot m^{-3}$)	3.20	3.06	2.93	2.62	1.86	1.05

5.1.3 Lattice constant and electron structure

The parameter of electron structure in EET includes the bond length and the amount of valence electrons. The change of lattice constant will influence the change of hybridized state and finally influence the change of electron structure, which indicates that there exists certain relationship between lattice constant and electron structure.

TFDC model describes electron structure in terms of atomic radius and electron density, and the electron density is dependent on the atomic radius. Wigner-Seitz atomic radius can be measured by X-ray diffraction. Consequently, atomic radius and lattice constant are the same thing.

In fact, both EET and TFDC use lattice constant to describe electron structure, so they can be correlated with each other via an atomic interaction volume[8].

In short, lattice constant can reveal the change of electron structure, and electron structure can also reveal the change of lattice constant.

5.2 Experiments of X-ray diffraction and results

5.2.1 Experimental methods

Bulk and powder specimens are prepared to measure the lattice constant of alloys. Among every group of specimens, one Al powder or Zn powder or mixed powder specimen in annealing state is simultaneously scanned to determine the accuracy of the diffractometer. The surface of bulk specimen is ground to be a plane for measuring the lattice constant. Due to the brittleness of Zn-5Al alloy at room temperature, its bulk specimen should be filed for obtaining the powder specimen with a mean grain size of 0.2 mm. Figure 5.1 is the X-ray diffraction spectrum of the mixed powder of 95 mass percent Zn and 5 mass percent Al in the annealing state by 350 ℃×2 h, and Table 5.4 is its diffraction data. The interplanar distance, the lattice constant and the atomic radius in the following tables are calculated in the unit of 10^{-10} m. Figures 5.2 to 5.4 are X-ray diffraction patterns of Zn-5Al alloy in rolled state, annealed and superplastically deformed state respectively, and Tables 5. 5 to 5. 7 are their diffraction data correspondingly. The comparison between the data of the present experiments and those from the literatures is shown in Table 5.8, which shows the effect of alloy states on the interplanar distance is at the order of 0. 001 and the diffractometer used in the experiments has an accuracy of 0.0001. X-ray diffraction experiments were carried out in the Materials Science Institute of Tsinghua University with the D/max-RB diffractometer.

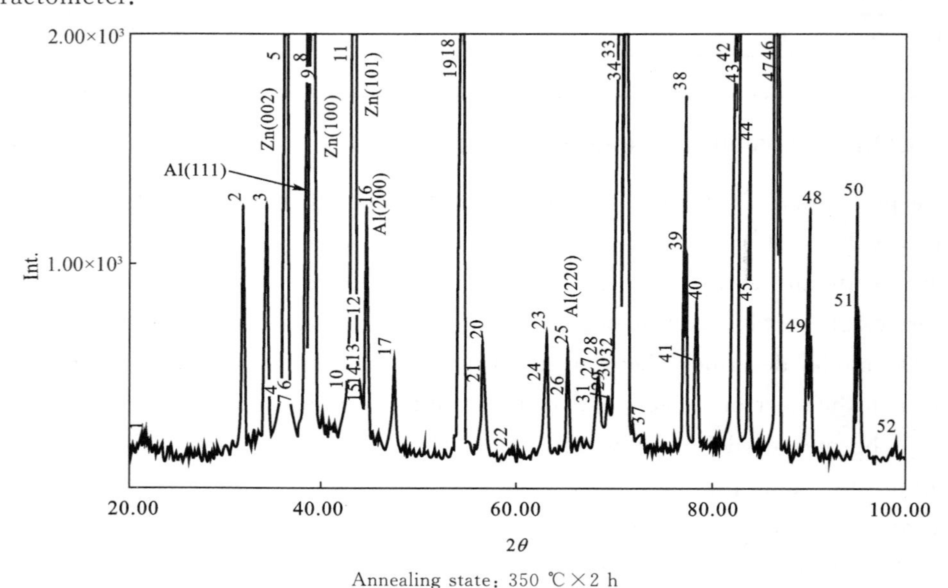

Annealing state: 350 ℃×2 h

Figure 5.1 X-ray diffraction pattern of Zn powder mixed with 5 mass percent Al powder.

Table 5.4 X-ray diffraction data for Zn powder mixed with 5 mass percent Al powder

No.	2θ	Int.	Width	D	I	Planes
1	21.400	243		4.1488	1	
2	31.800	1245	0.390	2.8117	3	
3	34.440	1245	0.420	2.6020	3	
4	35.640	397	0.240	2.5170	1	
5	36.300	15726	0.300	2.4728	38	Zn(002)

Continued

No.	2θ	Int.	Width	D	I	Planes
6	36.840	401	0.120	2.4378	1	
7	36.980	375	0.120	2.4289	1	
8	38.480	2781	0.270	2.3376	7	Al(111)
9	39.000	15397	0.300	2.3076	37	Zn(100)
10	42.520	431	0.210	2.1243	1	
11	43.240	41867	0.390	2.0906	100	Zn(101)
12	43.660	763	0.120	2.0175	2	
13	43.960	549	0.150	2.0580	1	
14	44.120	443	0.210	2.0509	1	
15	44.360	383	0.150	2.0404	1	
16	44.720	1236	0.240	2.0248	3	Al(200)
17	47.500	574	0.510	1.9088	1	
18	54.300	7257	0.240	1.6880	17	
19	54.400	5313	0.180	1.6851	13	
20	56.640	645	0.390	1.6237	2	
21	56.820	462	0.120	1.6190	1	
22	59.360	182		1.5556	0	
23	62.840	690	0.330	1.4776	2	
24	63.000	466	0.150	1.4742	1	
25	65.080	625	0.240	1.4320	1	Al(220)

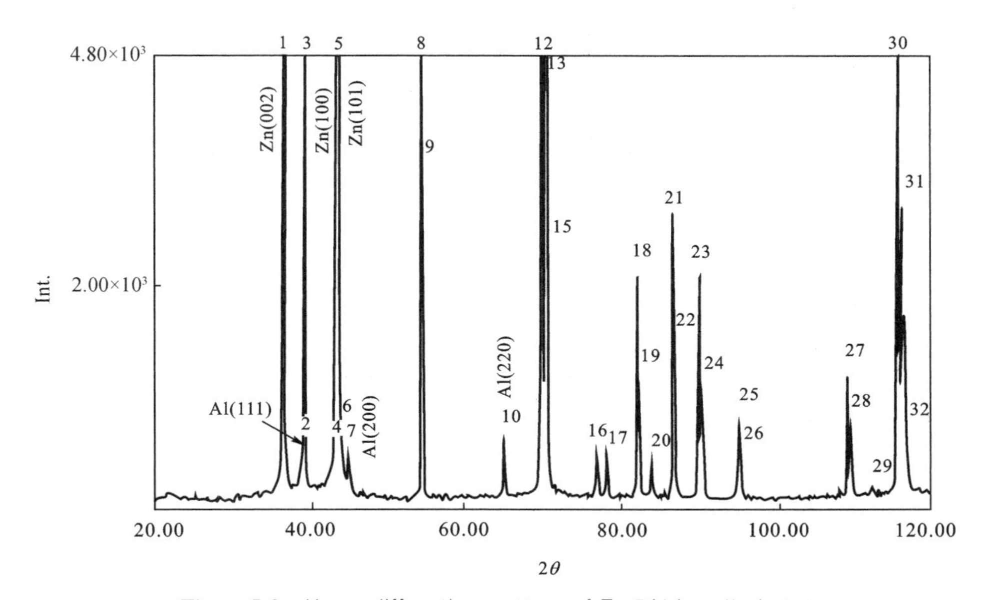

Figure 5.2 X-ray diffraction pattern of Zn-5Al in rolled state.

Table 5.5 X-ray diffraction data of Zn-5Al in rolled state

No.	2θ	Int.	Width	D	I	Planes
1	36.240	10809	0.300	2.4767	28	Zn(002)
2	38.460	656	0.270	2.3387	2	Al(111)
3	38.940	6988	0.300	2.3110	18	Zn(100)
4	42.580	546	0.150	2.1215	1	
5	43.180	38132	0.390	2.0934	100	Zn(101)
6	43.660	712	0.120	2.0715	2	
7	44.680	506	0.180	2.0265	1	Al(200)
8	54.280	5204	0.240	1.6886	14	
9	54.400	2997	0.120	1.6851	8	
10	65.100	660	0.330	1.4316	2	Al(220)
11	69.700	477	0.120	1.3480	1	
12	70.020	7572	0.240	1.3426	20	

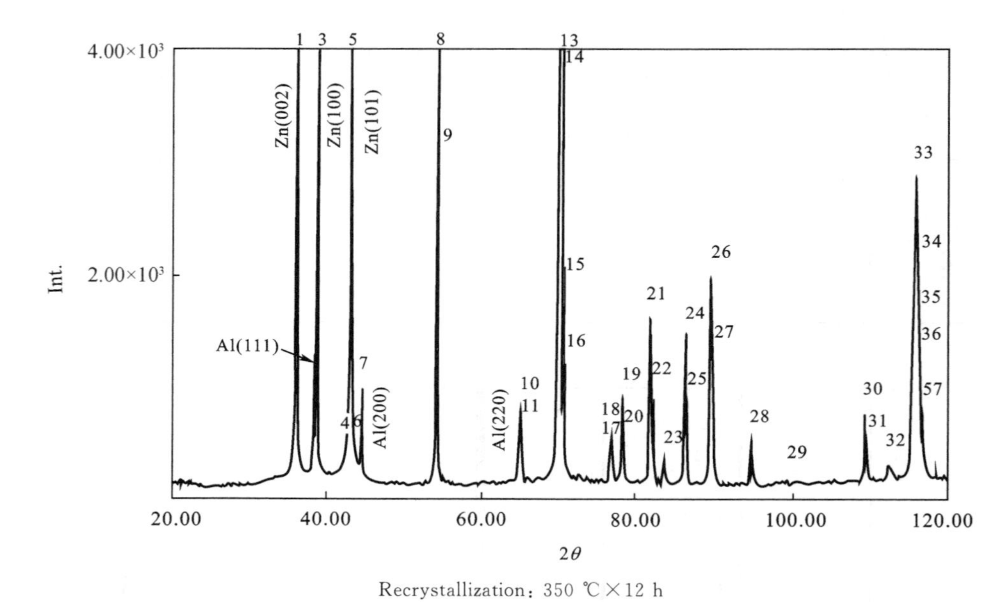

Figure 5.3 X-ray diffraction pattern of Zn-5Al in recrystallization state.

Table 5.6 X-ray diffraction data of Zn-5Al in recrystallization state

No.	2θ	Int.	Width	D	I	Planes
1	36.240	13052	0.300	2.4767	49	Zn(002)
2	38.480	1293	0.330	2.3376	5	Al(111)
3	38.980	5132	0.270	2.3087	19	Zn(100)
4	42.480	383	0.180	2.1262	1	
5	43.200	26754	0.360	2.0924	100	Zn(101)

Continued

No.	2θ	Int.	Width	D	I	Planes
6	43.820	459	0.120	2.0643	2	
7	44.740	993	0.390	2.0239	4	Al(200)
8	54.280	4370	0.240	1.6886	16	
9	54.380	3030	0.150	1.6857	11	
10	65.160	806	0.300	1.4305	3	Al(220)
11	65.260	608	0.120	1.4285	2	

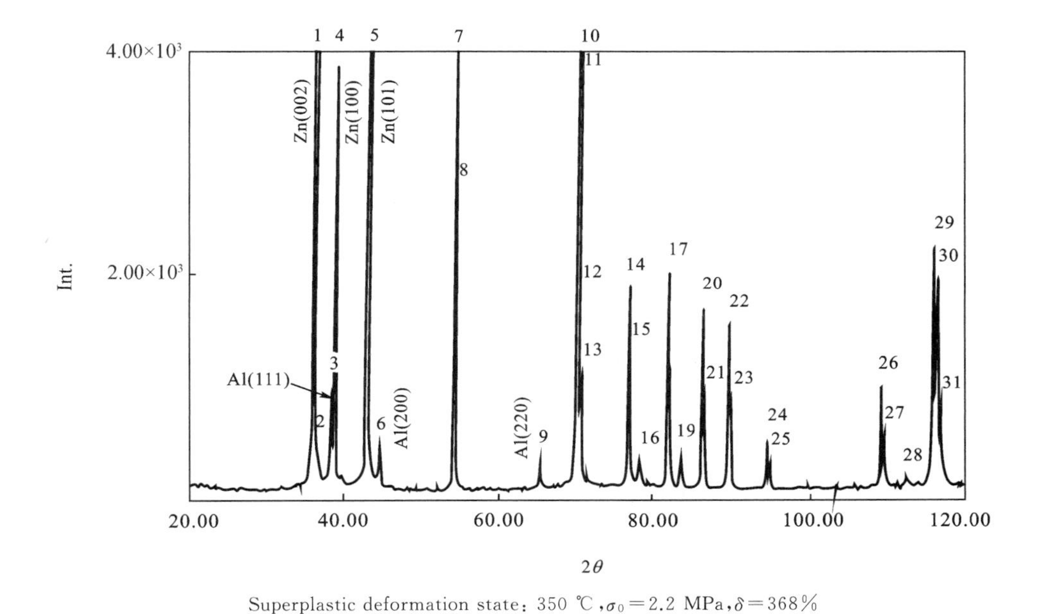

Superplastic deformation state: 350 ℃, σ_0=2.2 MPa, δ=368%

Figure 5.4 X-ray diffraction pattern of Zn-5Al in superplastic deformation state.

Table 5.7 X-ray diffraction data of Zn-5Al in superplastic deformation state

No.	2θ	Int.	Width	D	I	Planes
1	36.240	24748	0.330	2.4767	100	Zn(002)
2	36.700	486	0.120	2.4467	2	
3	38.540	956	0.270	2.3341	4	Al(111)
4	38.960	3849	0.300	2.3098	16	Zn(100)
5	43.200	19266	0.300	2.0924	78	Zn(101)
6	44.740	464	0.270	2.0239	2	Al(200)
7	54.280	4504	0.210	1.6886	18	
8	54.400	2728	0.120	1.6851	11	
9	65.180	371	0.180	1.4301	1	Al(220)
10	70.020	8074	0.240	1.3426	33	

Table 5.8 The comparison between the experimental and the literature's data

Plane	1992[51]	1951[52]	Experiment*	Δd**
Al(111)	2.3376	2.338	2.3376	±0.0000
Al(200)	2.0244	2.024	2.0248	±0.0004
Al(220)	1.4315	1.431	1.4320	±0.0005
Zn(002)	2.4735	2.473	2.4728	±0.0007
Zn(100)	2.3080	2.308	2.3076	±0.0004
Zn(101)	2.0915	2.091	2.0906	±0.0009

* Corresponding to Table 5.4.

** The difference between the literature's data in 1992 and the present experimental data.

5.2.2 Experimental results

The lattice constant of pure Al, a_0 is 4.0488×10^{-10} m; the lattice constants of pure Zn, a_0 is 2.6650×10^{-10} m and c_0 is 4.9470×10^{-10} m and c_0/a_0 is 1.8563[53].

For Zn-5Al alloy in rolled state, X-ray diffraction results of the bulk and the powder specimen are shown respectively in Table 5.9 and Table 5.10, where d is the interplanar distance of alloy phase and d_0 is the interplanar distance of pure metal.

Table 5.9 X-ray diffraction results of the bulk specimen for Zn-5Al alloy in rolled state

Phase	Diffraction	d	$d-d_0$
Al	(111)	2.3387	0.0011
Al	(200)	2.0265	0.0021
Al	(220)	—	
Zn	(002)	2.4767	0.0032
Zn	(100)	2.3110	0.0030
Zn	(101)	2.0934	0.0019

Al: $a=4.0518$, $a-a_0=0.003$.

Zn: $a=2.6685$; $a-a_0=0.0035$; $c=4.9535$; $c-c_0=0.0065$; $c/a=1.8562$; $c/a-c_0/a_0=0.0001$.

Table 5.10 X-ray diffraction results of the powder specimen for Zn-5Al alloy in rolled state

Phase	Diffraction	d	$d-d_0$
Al	(111)	2.3393	0.0017
Al	(200)	2.0257	0.0013
Al	(220)	—	
Zn	(002)	2.4761	0.0026
Zn	(100)	2.3104	0.0024
Zn	(101)	2.0924	0.0011

Al: $a=4.0516$; $a-a_0=0.003$.

Zn: $a=2.6678$; $a-a_0=0.003$; $c=4.9522$; $c-c_0=0.005$; $c/a=1.8563$; $c/a-c_0/a_0=0.0000$.

For Zn-5Al alloy in recrystallized state, X-ray diffraction results of the bulk and the powder specimen are shown respectively in Table 5.11 and Table 5.12, where d is the interplanar distance of alloy phase and d_0 is the interplanar distance of pure metal.

Table 5.11 X-ray diffraction results of the bulk specimen for Zn-5Al alloy in recrystallized state by 350 ℃ ×12 h

Phase	Diffraction	d	$d-d_0$
Al	(111)	2.3376	0.0000
Al	(200)	2.0239	−0.0005
Al	(220)	1.4305	−0.0010
Zn	(002)	2.4767	0.0032
Zn	(100)	2.3087	0.0007
Zn	(101)	2.0924	0.0009

Al: $a=4.0476$; $a-a_0=-0.001$.

Zn: $a=2.6658$; $a-a_0=0.0008$; $c=4.9534$; $c-c_0=0.006$; $c/a=1.8581$; $c/a-c_0/a_0=0.002$.

Table 5.12 X-ray diffraction results of the powder specimen for Zn-5Al alloy in recrystallized state by 350 ℃ ×12 h

Phase	Diffraction	d	$d-d_0$
Al	(111)	2.3364	−0.0012
Al	(200)	2.0235	−0.0009
Al	(220)	—	
Zn	(002)	2.4761	0.0026
Zn	(100)	2.3076	−0.0004
Zn	(101)	2.0924	0.0009

Al: $a=4.0469$; $a-a_0=-0.002$.

Zn: $a=2.6646$; $a-a_0=-0.0004$; $c=4.9522$; $c-c_0=0.005$; $c/a=1.8585$; $c/a-c_0/a_0=0.002$.

For Zn-5Al alloy in superplastically deformed state, X-ray diffraction results of the bulk and the powder specimen are shown respectively in Table 5.13 and Table 5.14, where d is the interplanar distance of alloy phase and d_0 is the interplanar distance of pure metal.

Table 5.13 X-ray diffraction results of the bulk specimen for Zn-5Al alloy in superplastically deformed state at the temperature of 350 ℃, under the stress of 2.2 MPa and by deformation of 368%

Phase	Diffraction	d	$d-d_0$
Al	(111)	2.3341	−0.0035
Al	(200)	2.0239	−0.0005
Al	(220)	1.4301	−0.0014
Zn	(002)	2.4767	0.0032
Zn	(100)	2.3098	0.0018
Zn	(101)	2.0924	0.0009

Al: $a=4.0452$; $a-a_0=-0.004$.

Zn: $a=2.6661$; $a-a_0=0.001$; $c=4.9534$; $c-c_0=0.006$; $c/a=1.8579$; $c/a-c_0/a_0=0.002$.

Table 5.14 X-ray diffraction results of the powder specimen for Zn-5Al alloy in superplastically deformed state at the temperature of 350 ℃, under the stress of 2.2 MPa and by deformation of 368%

Phase	Diffraction	d	$d-d_0$
Al	(111)	2.3364	−0.0012
Al	(200)	2.0222	−0.0022
Al	(220)	1.4301	−0.0014
Zn	(002)	2.4741	0.0006
Zn	(100)	2.3098	0.0018
Zn	(101)	2.0924	0.0009

Al: $a=4.0444$; $a-a_0=-0.004$.

Zn: $a=2.6671$; $a-a_0=0.002$; $c=4.9482$; $c-c_0=0.001$; $c/a=1.8553$; $c/a-c_0/a_0=-0.001$.

5.3 Explanation of changes of lattice constant

For the sake of clarity, the average results of the bulk and powder of X-ray diffraction in rolled state, recrystallized state and superplastically deformed state are put together with the density of the alloy in Table 5.15.

Table 5.15 The lattice constants of alloy phases, the cell strain and the density of Zn-5Al alloy

Alloy state	Lattice constants (nm)		Cell strain		Density ($g \cdot cm^{-3}$)
	Al	Zn	Al	Zn	
Rolled	0.40514	0.26676	0.00074	0.00100	6.8993
Recrystallized	0.40472	0.26668	−0.00042	0.00060	6.6506
Superplastically deformed	0.40448	0.26671	−0.00095	0.00080	5.8848

5.3.1 Changes of lattice constant in rolled alloy

It is shown in Table 5.15 that the lattice constants of rich-Al phase and rich-Zn phase in rolled state become larger, and the corresponding alloy density, 6.8993 g/cm^3, is also larger than that in recrystallized state, 6.6506 g/cm^3, and that in superplastically deformed state, 5.8848 g/cm^3. After being rolled, both the density of the alloy and the lattice constant of each phase increase.

It needs to explain why both the density and the lattice constant of the rolled alloy increase. When an alloy is rolled, its density will increase with the decreasing of defects such as pore, vacancy. However the increase of lattice constant can be explained by the anharmonic interaction between atoms, i. e., non-symmetry of attractive force and repulsion between atoms. The relationship of interaction force and distance between atoms is illustrated in Figure 5.5. With the same distance between atoms increased and decreased, the increase of repulsion between atoms is not equal to that of attractive force between atoms. In other words, under the same applied force, the distance contracted between atoms is different from that expanded between atoms. The extended distance is larger than the compressed one, as shown in Figure 5.6. After an alloy is rolled, the changes of interplanar distances of grains are different from one orientation to another, and some interplanar distances may become larger and others may become smaller. If texture in the alloy is neglected, the result estimated statistically is that the rolling process makes the interplanar distances increase. Therefore, the lattice constants of rolled alloy become larger.

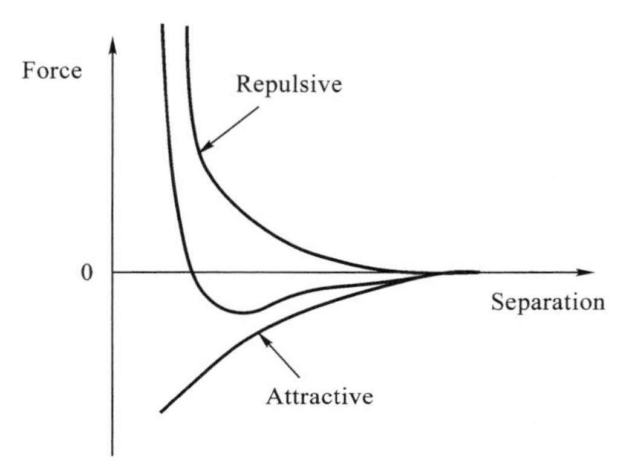

Figure 5.5 Schematic illustration of interaction force and distance between atoms.

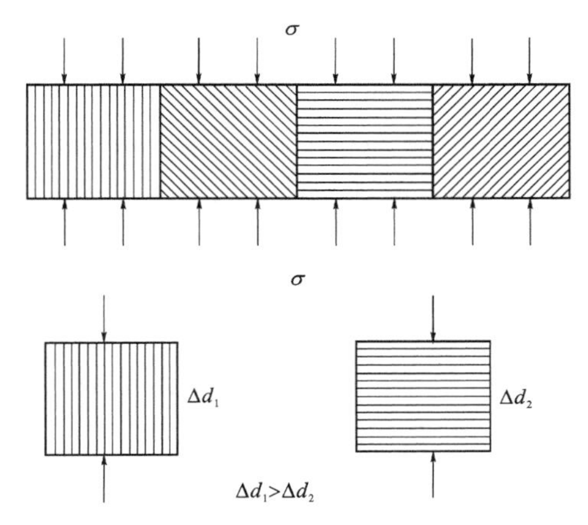

Figure 5.6 Schematic illustrations of changes of interplanar distances.

5.3.2 Changes of lattice constant in recrystallized and superplastically deformed alloys

For recrystallized Zn-5Al alloy, the lattice constant of rich-Al phase in the specimen is smaller than that of pure Al metal, and the lattice constant of rich-Zn phase in the alloy is larger than that of pure Zn metal.

For superplastically deformed Zn-5Al alloy, the lattice constant of rich-Al phase in the specimen is much more contracted, and that of rich-Zn phase is still expanded. The decrease of alloy's density is due to the fact that voids have existed during superplastic deformation.

The lattice constants of rich-Al and rich-Zn phases relate to the states of alloys, as shown in Table 5.15. According to close-packed lattice model, the lattice constant of crystal is related to the atomic radius, while the atomic radius depends on the electron distribution in the atom. Consequently, it can be believed that the lattice constant is determined by the interaction between the lattice and the electrons.

For rolled Zn-5Al alloy, its density is the largest, and the lattice of rich-Al phase and that of rich-Zn phase are respectively larger than that of the pure Al metal and pure Zn metal, which demonstrates that the crystal lattice plays the dominating role in the changes of the lattice constants. For recrystallized Zn-5Al alloy, its density becomes smaller, and the change of the lattice constant of rich-Al phase is contrary to that of rich-Zn phase, i.e., the lattice constant of rich-Al phase is smaller than that of the pure Al metal and the lattice constant of rich-Zn phase is larger than that of the pure Zn metal. For superplastically deformed Zn-5Al alloy, the density becomes much smaller than that in rolled and recrystallized state, the lattice constant of rich-Al phase becomes the smallest in three states and that of rich-Zn phase is still larger than that of pure Zn metal, which shows that role of the crystal lattice in the changes of the lattice constants becomes the smallest in three states.

For single-phase Al-Zn alloy, the lattice constant of Al phase decreases slightly due to the addition of Zn, because the atomic radius of Zn, 0.155 nm, is somewhat smaller than the atomic radius of Al, 0.158 nm[48]. However in the two-phase Zn-Al alloy, the change of the lattice constants of the rich-Al phase is not only dependent on the rich-Zn phase, but also dependent on the interface between the rich-Al phase and the rich-Zn phase.

It can be seen from Al-Zn phase diagram[4] that the solvency of Zn in Al is the same for Zn-5Al and Zn-10Al alloys, but the amount of interfaces is quite different, which leads to different changes of the lattice constant of rich-Al phase and rich-Zn phase in the two alloys, as shown in Table 5.16.

The phenomenon that the rich-Al phase contracts and the rich-Zn phase expands for Zn-5Al eutectic alloy in recrystallized and superplastically deformed states can be explained qualitatively by TFDC model.

Table 5.16 The lattice constants of alloy phases, the cell strain and the density of Zn-10Al alloy

Alloy state	Lattice constants (nm)		Cell strain		Density (g · cm^{-3})
	Al	Zn	Al	Zn	
Rolled	0.40517	0.26670	0.00072	0.00076	6.2808
Recrystallized	0.40502	0.26672	0.00042	0.00082	6.2192
Superplastically deformed	0.40494	0.26664	0.00015	0.00052	6.1321

Note: Recrystallization, 350 ℃ × 12 h; superplastic deformation, 350 ℃; deformation stress, 2.2 MPa; elongation: 625%.

5.3.3 Explanation of changes of lattice constant by TFDC model

A simple two-atom model is proposed according to TFDC model for explaining the changes of lattice constant. In this model, the interface of the rich-Al phase and the rich-Zn phase is considered as the interface of an Al atom and a Zn atom. Although the orientation of the interface and the solvency of Zn in Al are not taken into account, the two-atom model can qualitatively explain the phenomenon that the lattice constant of the rich-Al phase shortens and the lattice constant of the rich-Zn phase lengthens for Zn-5Al eutectic alloy in recrystallized and superplastically deformed states.

TFDC model indicates that the boundary condition between the atoms in a solid is that the density of the electrons between the atoms at the interface is equal[50]. This condition is the continuity of wave function in the quantum mechanics. The radii of the atoms and the densities of the electrons at the radius for pure Al metal and Zn metal given by TFDC model are

$$r_{Al}=1.58\times10^{-8}\ \text{cm} \tag{5-1}$$

$$r_{Zn}=1.55\times10^{-8}\ \text{cm} \tag{5-2}$$

$$n_{Al}=1.05\times10^{23}\ \text{cm}^{-3} \tag{5-3}$$

$$n_{Zn}=1.86\times10^{23}\ \text{cm}^{-3} \tag{5-4}$$

When the interface is formed between Al atoms and Zn atoms in the alloy, the boundary condition requires that the densities of the electrons between the atoms in the interface should be equal. As a result, the atoms with low density of the electrons are contracted and the atoms with high density of the electrons are expanded in order to ensure the boundary condition. In fact, the radius of the atom is the same as the lattice constant for the crystal solid. It is the electron effect that leads to the contraction of the lattice constant of rich-Al phase and the expansion of the lattice constant of rich-Zn phase in Zn-5Al alloy.

Based on the above analysis, using the two-atom model, the strain acting on the rich-Al phase and the rich-Zn phase can be calculated. The mean electron density on the interface obtained by Equation (5-3) and Equation (5-4) is

$$n=1.455\times10^{23}\ \text{cm}^{-3} \tag{5-5}$$

Suppose that the valence electrons of the metallic bond in the pure Al and the pure Zn metal are the same as that of the metallic bond in the alloy, and these valence electrons can be described with Equations (5-3) and (5-4). In order to satisfy Equation (5-5) for the atoms on the interface, both the electron density of the Al atom and that of the Zn atom should be n. In the case that the valence electrons are constant, in order to change the electron densities of the Al atom and the Zn atom into n, the radii of the atoms must change as follows

$$n_{Al}V_{Al}=nV'_{Al} \tag{5-6}$$

$$n_{Zn}V_{Zn}=nV'_{Zn} \tag{5-7}$$

where V and V' are the atomic volume before and after the change respectively.

The atomic radii can be obtained from Equations (5-1) to (5-7), namely

$$r'_{Al}=1.4172\times10^{-8}\ \text{cm} \tag{5-8}$$

$$r'_{Zn}=1.6854\times10^{-8}\ \text{cm} \tag{5-9}$$

Equations (5-8) and (5-9) being compared with Equations (5-1) and (5-2), it can be seen that the radius of the atom of the Al decreases and that of the Zn increases. The strains of the crystal cells produced by the change of the radii of the atoms are as follows

$$\varepsilon_{Al}=-0.1030 \tag{5-10}$$

$$\varepsilon_{Zn}=0.0852 \tag{5-11}$$

where the negative sign corresponds to compression and the positive sign corresponds to expansion.

In fact, the rich-Al phase and the rich-Zn phase in Zn-5Al alloy are the solid solution. At room temperature, the solvency of Al in Zn is nearly zero and that of Zn in Al is 0.82 atom percent[53]. At present, the solvency is not important and the most important factors are the structure and the amount of the interfaces. For example, the solvency of Zn in Al in Zn-5Al alloys is the same as that in Zn-10Al alloys, but the experimental results of these two alloys concerning the changes of the lattice constants are quite different, because the structure and the amount of the interface between Al phase and Zn phase in Zn-5Al alloy are different from those in Zn-10Al alloy.

It should be pointed out that the theoretical values are much larger than the experimental values due to the approximation and the simplified calculation of Equations (5-6) and (5-7).

5.4 Appearance of electron effect in superplasticity

What role do the electrons play in the superplastic deformation? To answer this problem involves which is larger during superplastic deformation, the effect of electrons or the effect of the crystal lattice. The following is a discussion about this problem. First, the two examples exhibiting obvious electrons effects are introduced. Then, the role of the electron in superplastic deformation is discussed.

5.4.1 Examples about interaction between electron and lattice

(1) Stack fault energy

Electrons are main contributors to stack fault energy of the crystal[54], because stack faults do not result in any distortion of the crystal lattice. Using electron density functional theory, Wang Chongyu[55] deduced that stack fault energy of Ni metal is $152.8\times10^{-7}\ \text{J/cm}^2$ (The experimental result from a literature is $150\times10^{-7}\ \text{J/cm}^2$). Stack fault is a kind of electron effect, but it behaves as the stacking defects of the crystal.

(2) Vegard Law

For continuous solid solution, the relationship of the lattice constants and the concentration of the solute given by Vegard in 1921 is as follows

$$a=(1-x)a_1+xa_2 \tag{5-12}$$

where a, a_1 and a_2 are the lattice constants corresponding to the solid solution, the solvent and the solute respectively, and x is the concentration of the solute.

In most cases, Vegard Law does not agree with the experimental results. If the atoms are rigid sphere, Vegard Law has no difficulty in describing the lattice constant of solid solution. In fact, the lattice constant of the solid solution is not only dependent on the lattice constant of the component, but also dependent on the electron factors of the component, as discussed in Section 1.3.

5.4.2 Investigation of interaction between electron and lattice in interface

From microscopic view, the contraction of the atoms in α phase and the expansion of the atoms in β phase correspond to α phase bearing the compressive stress and β phase bearing the extension stress. The microscopic interaction energy between Al atoms and Zn atoms at interface is the macroscopic interface energy of α/β. If β phase is thought of as a kind of medium, the surface energy of α phase in β phase may be written as $\gamma_{\alpha/\beta}$. According to Laplace equation[56], the macroscopic pressure acting on α phase is given by

$$\tau=\gamma_{\alpha/\beta}\left(\frac{1}{\rho_1}+\frac{1}{\rho_2}\right) \tag{5-13}$$

where ρ_1 and ρ_2 are two local principal radius of curvature of α phase.

Due to the bar shape of α phase with the length far larger than width and thickness, Equation (5-13) can be simplified as

$$\tau=\frac{\gamma_{\alpha/\beta}}{\rho} \tag{5-14}$$

where ρ is the radius of curvature of α phase, $\gamma_{\alpha/\beta}$ is the surface energy of α phase in the medium of β phase, τ is the compressed stress born by α phase.

When α phase is in straight line, $\rho=\infty$ and $\tau=0$, which indicates that α bar does not endure any force and is in a stable structure. It is shown by TEM that the α bars in cast state are all straight.

During superplastic deformation, α bar unavoidably becomes bent, which means that radius of curvature of α bar turns from infinite to finite and α phase begins to bear certain macroscopic stress. The smaller the radius of curvature is, the larger the stress is. In order to decrease the stress, the bent position of α phase will become thicker and thicker until the dumbbell-shaped α phase occurs. Figure 5.7 is TEM images of Zn-5Al alloy with a deformation of 306%. In the figure, there are many dumbbell-shaped α phases.

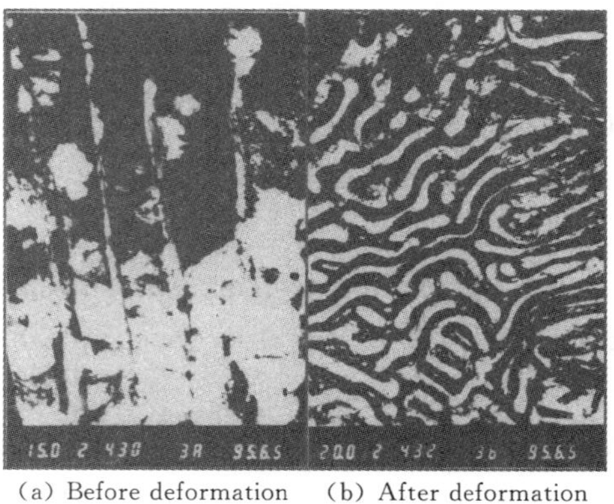

(a) Before deformation (b) After deformation

350 ℃, 1.0×10^{-3}/s, $\delta=306\%$

Figure 5.7 TEM images of Zn-5Al alloy.

5.4.3 Role of interface stress of α phase in superplastic deformation

In Chapter 4, the circumstances that the angle between the interface α/β and the tensile direction is advantageous to sliding are discussed. In the case that the angle between the interface α/β and the tensile direction is disadvantageous to sliding, under the action of the load, α phase will become bent and the stress born by α phase increases. In order to decrease the stress, α phase will increase its radius of curvature by means of the migration of Al atoms to the places with smaller radius of curvature. As a result, the α bar at the straight position near its bent parts becomes thinner and thinner until it breaks finally. Therefore, under the action of load and interface tension, the long α phases disadvantaging sliding are changed into short ones advantaging superplasticity.

5.5 Summary

The density and the lattice constant of Zn-5Al alloy in rolled state simultaneously increase. Increase of density is due to decrease of pores or voids in the rolled alloy, and increase of the lattice constant is due to the anharmonic interaction of atoms with the change of distance between atoms. In recrystallized state, the lattice constant of the Al phase becomes smaller. After superplastic deformation, the lattice constant of the Al phase becomes even smaller than that of pure Al metal, while the lattice constant of Zn phase is still larger than that of the pure Zn metal. The effect is caused by the fact that, when the interface between Al phase and Zn phase in the alloy is formed, the Al atoms with lower electron density contract and the Zn atoms with higher electron density expand, in order to satisfy the equality of electron density on the interface. It follows that the Al phase bears a compressed stress and the Zn phase bears a tensile stress.

The interaction energy of Al atoms and Zn atoms in interface α/β on the microscopic scale is just $\gamma_{\alpha/\beta}$, the energy of interface α/β on the macroscopic scale. $\gamma_{\alpha/\beta}$ can also be considered as surface energy of α phase in the medium of β phase. When α phase is straight, the radius of curvature is infinite, the α bar does not bear any force. Under the action of load, α phase will become bent and the stress born by α phase increases. In order to decrease the stress, α phase increases its radius of curvature by means of the migration of Al atoms to the places with smaller radius of curvature. As a result, the α phase develops into dumbbell-shape. In this way, under the action of the load and interface tension, long α phases disadvantaging sliding are changed into several short ones advantaging sliding.

Chapter 6

Effect of Electric Field on Diffusion Dissolution Zone

The mechanism of the superplastic deformation in Zn-5Al eutectic alloy is the interface sliding controlled by a diffusion dissolution zone, and the broadening of the diffusion dissolution zone is disadvantageous to the interface sliding. The Zn-5Al alloy deformed superplastically in an external electric field displays a negative electroplasticity that the stress increases and the strain rate decreases and the voids develop.

In this chapter, firstly, the positive electroplastic effect is introduced. Then the experimental method about the negative electroplastic effect is described. Finally the mechanism's effects are dealt with.

6.1 Positive electroplastic effect

The effect of electric fields and currents on deformation (sometimes, indicating dislocation mobility) in metals has been termed electroplasticity[57-58]. Electroplasticity is divided into the electroplasticity employed electric currents and that employed electric fields according to experimental methods. Sometimes, the effect of electric currents on dislocation's movement or atomic mobility is also called electroplasticity.

So far, the reported electroplastic effects regardless of employing electric fields or currents, all exhibit a positive effect that the plastic or superplastic deformation stresses were decreased and the elongation and the strain rate were increased. It was thought that the effect caused by the electric factors affects the atomic migration and the dislocation movement. Opposite to the above is the negative electroplasticity that the deformation stresses are increased and the elongation and the strain rate are decreased by the electric factors.

6.1.1 Electroplasticity employed electric currents

The electroplasticity employed electric currents is usually tested above $0.5T_m$ or below $0.5T_m$, where T_m is melt point of specimen. The former was seldom reported and the latter is reviewed as follows. The electric current employed in electroplasticity, generally, is high-density electric current pulses(10^3-10^7 A/cm^2 for 100 μs and 2 Hz).

Conrad[58] thought that the effect of electric currents on deformation is due to an interaction between conduction electrons and dislocations, and the interaction is called electron wind force. Based on the consideration of the specific dislocation resistivity, the electron wind force per unit length of dislocation is given by

$$F_{ew}=\rho_D e n_e j/N_D \tag{6-1}$$

where ρ_D/N_D is the specific dislocation resistivity, e is the electron charge, n_e is the electron density, j is the current density.

While based on quantum mechanics consideration of the interaction between conduction electrons and dislocations, F_{ew} is also given by

$$F_{ew}=\alpha b P_F(j/e-n_e v_D) \tag{6-2}$$

where α is a constant between 0.1 and 1.0, b is the Burgers vector, P_F is the Fermi momentum and v_D is the dislocation speed.

Knowing the electron wind force, one can obtain the electron wind push constant by

$$K_{ew}=F_{ew}/j \tag{6-3}$$

The electron wind push coefficient is defined by

$$B_{ew}=K_{ew} e n_e F_{ew}/v_e \tag{6-4}$$

where v_e is the electron speed.

For the face-centered cubic metals copper, silver and gold with $\alpha=0.25$, Equation (6-1) yields $B_{ew}=10^{-9}$ N · s/cm^2, whereas Equation (6-2) gives $B_{ew}=5\times10^{-10}$ N · s/cm^2.

Conrad and his co-workers finished a quantity of researches about electroplasticity with metals (Ag, Al, Cu, Ni, Nb, Fe, W and Ti) varying in purity from 99.9 to 99.999 mass percent. An important result was obtained by Conrad as follows

$$\dot{\varepsilon}_j/\dot{\varepsilon}_{j=0}=(j/j_c)^p \tag{6-5}$$

where $\dot{\varepsilon}_j$ is the strain rate with the current pulse, $\dot{\varepsilon}_{j=0}=0$ is the strain rate prior to the pulse, j is the current density(about 10^5 A/cm^2), j_c is the critical current density ($10^{-3}-10^{-4}$ A/cm^2), and p is an exponent of about 3.

For the various metals, j_c was found to increase with electron density, viz.,

$$j_c= C n_e^q \tag{6-6}$$

where q is about 2/3 at 300 K and 2/5 at 77 K, and C is a constant which decreases with temperature.

Conrad gave the following expressions for the resolved shear rate

$$\dot{\gamma}=\dot{\gamma}_0\exp\left[\frac{-(\Delta H^*-A^* b\tau^*)}{kT}\right] \tag{6-7}$$

where γ_0 is the pre-exponential factor including the entropy of activation, ΔH^* is the activation enthalpy, A^* is the activation area, b is the Burgers vector, τ^* is the resolved effective shear stress and kT has the usual significance.

$\dot{\gamma}_0$ is defined by

$$\dot{\gamma}_0 = N_{Dm} bs\nu^* \exp(\Delta S^* / k) \tag{6-8}$$

where N_{Dm} is the mobile dislocation density, b is the Burgers vector, s is the average distance that the dislocation segment has moved per successful thermal fluctuation, ν^* is the frequency of vibration of the dislocation segment and ΔS^* is the entropy of activation.

Conrad gave another equation

$$\ln\left(\frac{\dot{\gamma}_j}{\dot{\gamma}}\right) = \ln\left(\frac{\dot{\gamma}_{0j}}{\dot{\gamma}_0}\right) - \frac{\delta \Delta H^*}{kT} \tag{6-9}$$

The subscript j indicates the value with the current on, omission of this subscript indicates the value prior to the application of the current. The change in activation enthalpy is defined as

$$\delta \Delta H^* = \Delta U^* - p\Delta V - A^* b\tau^* - A^* F_{ew} \tag{6-10}$$

where ΔU^* is the activation internal energy, p is the pressure and ΔV is the change in volume.

For polycrystals, $\dot{\gamma} = M\dot{\varepsilon}$ and $\tau^* = \sigma^*/M$, where M (about 3) is the Taylor orientation factor, $\dot{\varepsilon}$ and σ^* are the true uniaxial strain rate and stress respectively.

In 1995, Conrad further researched the electroplasticity employed electric currents and it was thought that the physical mechanism's effect of the current on $\dot{\gamma}_0$ is not clear. It was very possible that the effect lie in the current affecting the frequency of vibration of the dislocation, and in fact, the essence of the problem was probably the interaction between electrons and phonons[59].

6.1.2 Electroplasticity employed electric fields

Electroplasticity employed electric fields was reported as early as in 1973[58], and the result was that electric fields increased the strain rate of Cu and Co. Conrad reported an electroplastic effect in 1989, as shown in Figure 6.1, that electric fields can decrease the stress by 15% during the superplastic deformation for 7475Al[60]. Afterwards, the researches in China reported similar results.

There were a few researches about electroplasticity employed electric fields, and so far any theory about it has not been reported. At the temperature of $0.5T_m$, the leakage current in specimen is only 10 nA/mm^2. Therefore, the theory about the electroplasticity employed electric current seems to be ineffective in the electroplasticity employed electric fields.

The first problem encountered in trying to explain electroplasticity employed electric fields is the depth of electric fields effect. For conductive metals, the effect of the electric fields only lie in the surface of specimen, i.e., induced charge on the surface of specimen, according to classical electromagnetics. However, Conrad accounted for

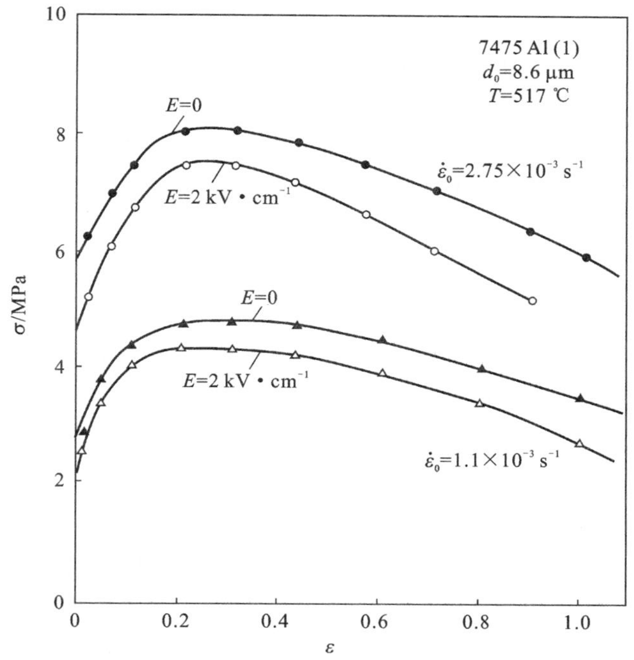

Figure 6.1 Effect of electric fields on stress vs. strain curve[60].

the fact that electric effects extended to the center of specimens with 1.8 mm in thickness. Starting with the relation $X=(Dt)^{1/2}$ where X is the diffusion distance, D is the diffusion coefficient and t is the time of the test, and taking the values $X=0.09$ cm and $D=0.2\exp\left(-\frac{11500}{RT}\right)$ for vacancy diffusion in Al, one obtains $t=56$ s. Considering the employed strain rates, this time is short enough that any changes in vacancy concentration at the surface could conceivably diffuse into the interior and thereby influence its mechanical behavior.

An alternative explanation for the fact that the electric field extends to the center of relatively thick specimens was cited by Conrad. It suggested the possibility of an uneven electron density at the interfaces between phases and at grain boundaries. The reduction in cavities produced by the field might be a reflection of this, since the cavitation occurred mainly along grain boundaries and at interfaces between phases.

The author's experiments about electroplasticity were finished in Conrad's laboratory and the result is different from that done by Conrad and his co-workers. Their results for Al alloy, Ti alloy and steel, etc. are that the electric fields decrease the deformation stress by 15%. However, the present result for Zn-Al alloy is that the electric fields decrease the strain rate by an order of magnitude.

6.2 Negative electroplasticity effect

In this section, the experimental equipment, experimental materials, experimental method and the reappearance of experiments about negative electroplasticity will be introduced.

6.2.1 Experimental equipment

To decrease the effect of electric fields on the thermocouple and the electron record system, a simple and reliable self-made instrument was used in the experiment. The temperature is measured with high-temperature mercury thermometer, and controlled by a voltage modulator with an accuracy of 2 ℃. The loads are added or subtracted by hanging or removing the weights with an accuracy of 1 g. The displacement is recorded with *X*-*t* recorder with an accuracy of 1 s and 0.5 mm for time and distance respectively.

To employ electric fields, the extended grips were made from Lawa ceramics, and two electrodes were installed outside the ceramic grips, in which one electrode is connected to the anode and the other is connected to the negative pole of the power, as shown in Figure 6.2. The distance between electrodes is 16 mm, and in the case that the voltage is 4.2 kV, the electric field intensity on the surface of specimen is 0.3 kV/mm.

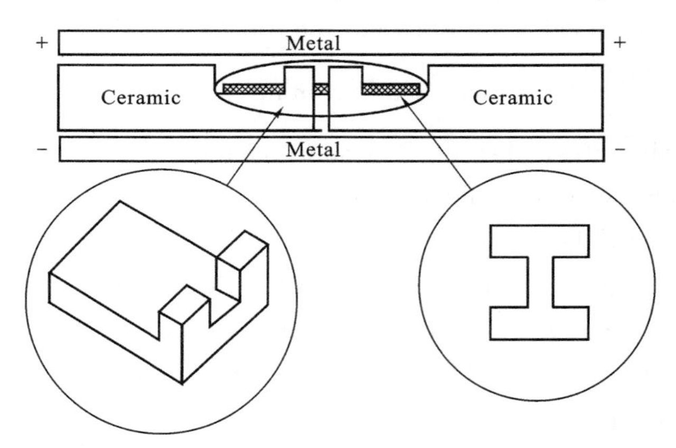

Figure 6.2 Extension grips and specimen arrangement.

6.2.2 Experimental materials and methods

Zn-5Al eutectic alloys are used in the experiments, and the specimens parallel to the rolling direction with gauge section 13 mm long and 6 mm wide are machined from rolled sheet of 2 mm thickness.

The surfaces of specimens are of two types: ground surface by the grind paper with grain of grade 600 and polished surface by Al_2O_3 with the grain of 1 μm. The states of the specimens also are of two types: recrystallized state and rolled state.

Electroplasticity in the former is too small, and that in rolled state will be discussed in this chapter.

6.2.3 Reappearance of experiments

The superplastic effect is very sensitive to the structures of materials, and especially so is the electroplastic effect. Since Conrad and his co-workers published the first paper about the electroplastic effect during the superplastic deformation in electric fields in 1989, some scholars have doubted their results.

Before the experiments about electroplasticity began, the author had inspected Conrad' s experimental methods and experimental equipment in his laboratory. For obtaining obvious and reliable results, some works are improved. Under the same condition of experiment, two specimens are tested, one with electric field and the other without electric field. The two specimens are machined from a piece of sheet. Under every condition, the experiments are repeated three times, and results are shown in Figure 6.3.

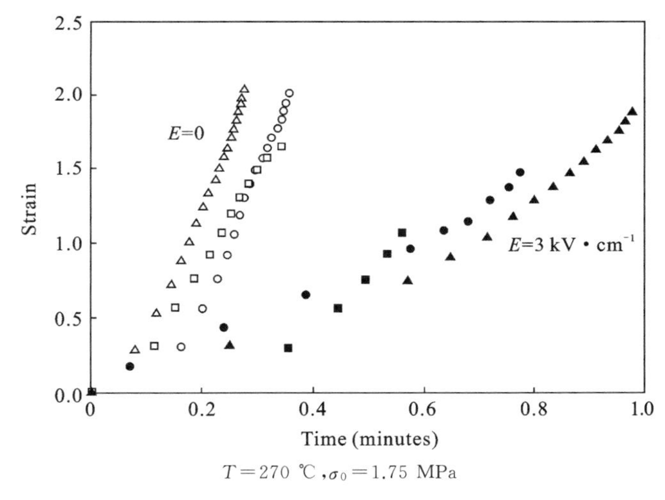

$T=270$ ℃, $\sigma_0=1.75$ MPa

Figure 6.3 Reappearance of electroplasticity in Zn-5Al alloy.

6.3 Experiments and analysis

According to the experimental value of the eutectoid temperature in Chapter 4, the experimental temperature can be divided into three zones: low temperature zone(<285 ℃), high temperature zone(>285 ℃) and critical temperature zone(approximately 285 ℃).

6.3.1 Effect in low temperature zone(<285 ℃)

It needs a few days or a decade of days that a specimen is tested with the constant load method under low initial stress($\sigma_0 = 1$ MPa) and at low temperature. Therefore, only the initial stage of deformation is studied at low temperature. The mean strain rates of the specimens deformed by 5% are shown in Table 6.1.

Table 6.1 The mean strain rates of Zn-5Al specimens deformed by 5%.

Electric field(kV/mm)	220 ℃	240 ℃	260 ℃
0.0	4.47×10^{-6}/s	6.87×10^{-6}/s	2.01×10^{-5}/s
0.3	3.38×10^{-6}/s	5.10×10^{-6}/s	1.13×10^{-5}/s

It is shown in Table 6.1 that the electric field makes the strain rate of the superplastic deformation decrease slightly. The data in Table 6.1 are recorded after the relaxation of 5 minutes. The strain rates may be thought of as constant under low strain at low temperature.

The relationship between the activation and the strain rate may be represented as

$$\dot{\varepsilon} G^{n-1} T = A\exp\left(-\frac{Q}{RT}\right) \tag{6-11}$$

where G is the shear modulus, R is the gas constant, Q is the activation energy, T is the absolute temperature, $\dot{\varepsilon}$ is the strain rate, n is the stress exponent usually taking 2 without electric field and 2.4 with electric field(see Equation(6-13)), A is a constant.

The curve of $\ln(\dot{\varepsilon} G^{n-1} T)$ versus $\frac{1}{T}$ in Figure 6.4 is plotted according to the data from Table 6.1. It is seen in Figure 6.4 that the slopes of the curves without and with electric field have little difference, and the activation energies calculated from the curves are 88.9 kJ/mol without electric field and 72.3 kJ/mol with electric field respectively. In view of the experimental error (<10%), it is believed that the electric field has no effect on the activation energy in the low temperature zone.

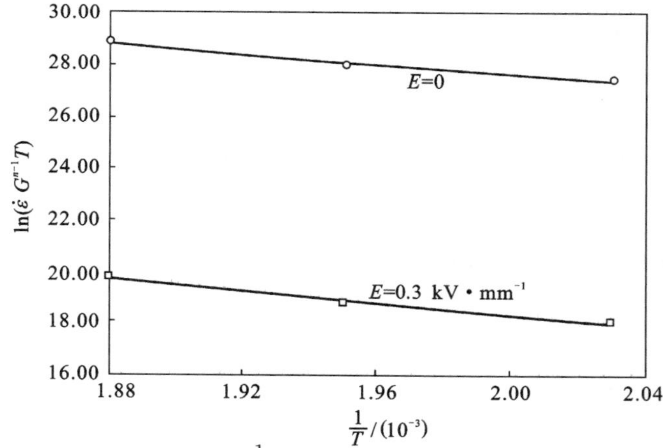

Figure 6.4 $\ln(\dot{\varepsilon} G^{n-1} T)$ vs. $\frac{1}{T}$ curve of Zn-5Al alloy at low temperature.

6.3.2 Effect in high temperature zone(>285 ℃)

Table 6.2 is the experimental results at high temperature under constant load ($\sigma_0 = 1$ MPa). The strain rates listed in the table all correspond to the strain of 0.4 and the same stress.

Table 6.2 The strain rate of Zn-5Al at the strain of 0.4

Electric field(kV/mm)	300 ℃	320 ℃	340 ℃
0.0	9.09×10^{-4}/s	2.35×10^{-3}/s	4.50×10^{-3}/s
0.35	6.94×10^{-4}/s	1.59×10^{-3}/s	4.10×10^{-3}/s

Based on the data in Table 6.2, the activation energy deduced from the gradient of the $\ln(\dot{\varepsilon} G^{n-1} T)$ versus $\frac{1}{T}$ curve according to Equation (6-11) is 99 kJ/mol (without electric field) and 132 kJ/mol (with electric field) respectively. Figure 6.5 shows the plot of $-\ln\dot{\varepsilon}$ versus $1/T$ from Table 6.2. The activation energy calculated from the gradient of the curve is 105.1 kJ/mol (without electric field) and 131.5 kJ/mol (with electric field). It can be seen that the results from these two methods are basically in agreement. The electric field has certain effect on the activation energy in the high temperature zone.

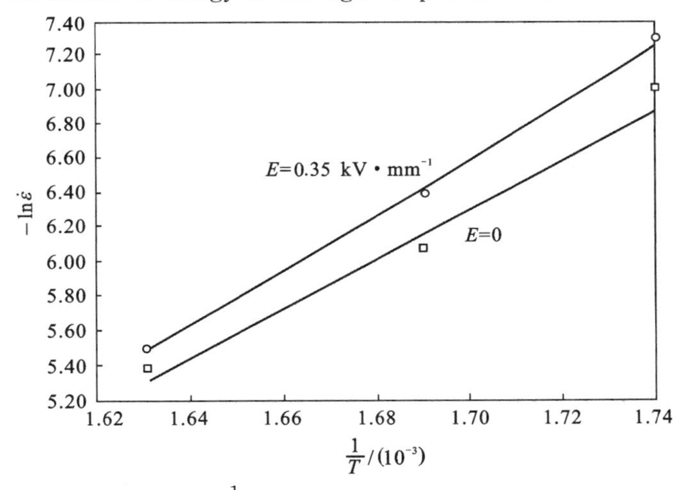

Figure 6.5 $-\ln\dot{\varepsilon}$ vs. $\frac{1}{T}$ curve of Zn-5Al alloy at high temperature.

Zn-22Al is two-phase alloy and three kinds of interfaces exist in the alloy. It is reported by the literature [61-63] that the activation energy of the interface sliding is 109 kJ/mol, 93 kJ/mol and 63 kJ/mol for α/α, α/β and β/β respectively. The activation energy of self-diffusion is 142 kJ/mol for Al and 93 kJ/mol for Zn[4]. For Zn-5Al alloy, in the high temperature zone, it seems that all these aspects are involved. Hence the complex circumstances need further studies.

6.3.3 Effect in critical temperature zone (approximately 285 ℃)

At the temperature ranging between 270 ℃ and 290 ℃, negative electroplastic effect is very pronounced. Figure 6.6 is the tensile result of rolled Zn-5Al alloy at 270 ℃ under constant stress of 1.0 MPa. As can be seen, the role of electric field seems to affect the strain rate. Figure 6.7 is the strain rate versus the strain curve corresponding to Figure 6.6.

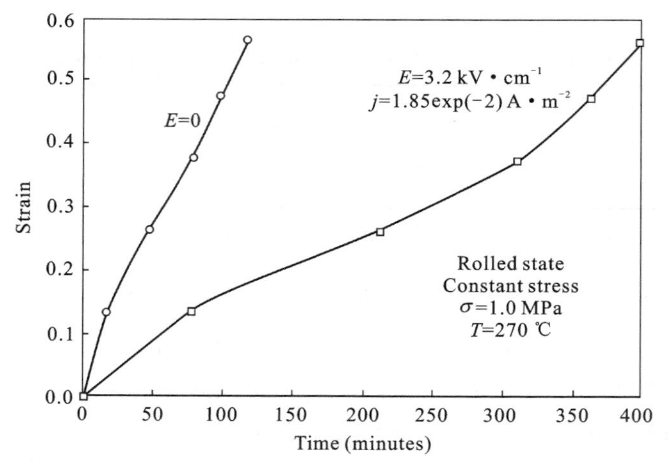

Figure 6.6 Tensile test of Zn-5Al alloy at 270 ℃ under constant stress.

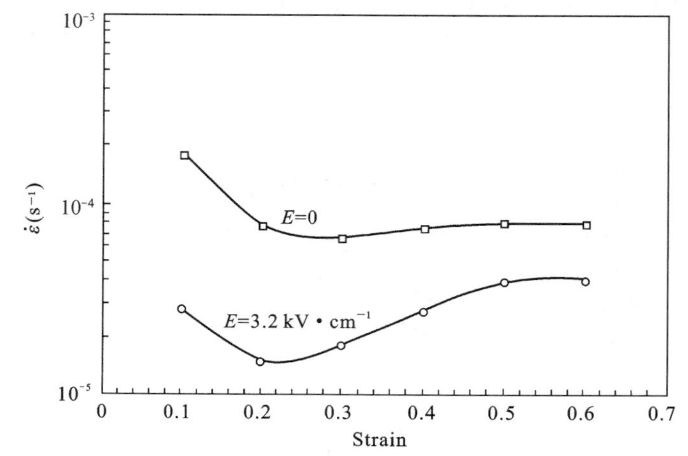

Figure 6.7 Effect of electric field on strain rate.

For the negative electroplastic effect, the strain rate can be described using the following equation

$$\delta \lg \dot{\varepsilon} = \lg \dot{\varepsilon}_{w/o} - \lg \dot{\varepsilon}_{w} \tag{6-12}$$

where $\dot{\varepsilon}_{w/o}$ is the strain rate without electric field and $\dot{\varepsilon}_{w}$ is the strain rate with electric field.

Figure 6.8 shows negative electroplastic effect as a function of the strain rate according to Equation (6-12). Figure 6.9 shows the stress exponent n as a function of the strain rate at 270 ℃ under constant stress. In the case that ε is between 0.2 and 0.6, the following relation can be obtained.

$$n_e / n_o = 1.2 \tag{6-13}$$

where n_e is the stress exponent with electric field and n_0 is that without electric field.

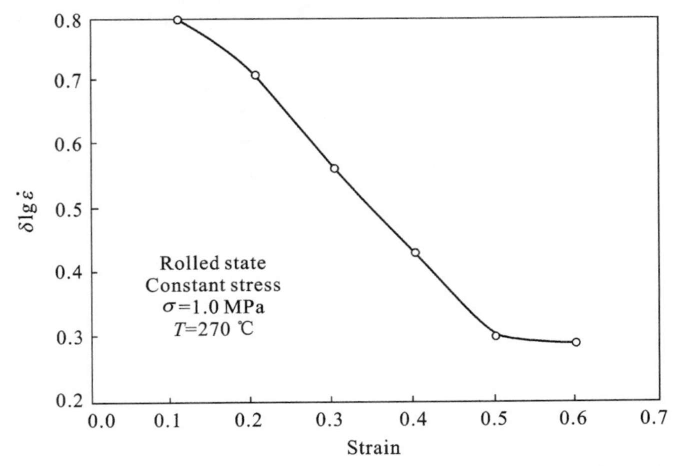

Figure 6.8 The $\delta \lg \dot{\varepsilon}$ versus ε curve of negative electroplasticity.

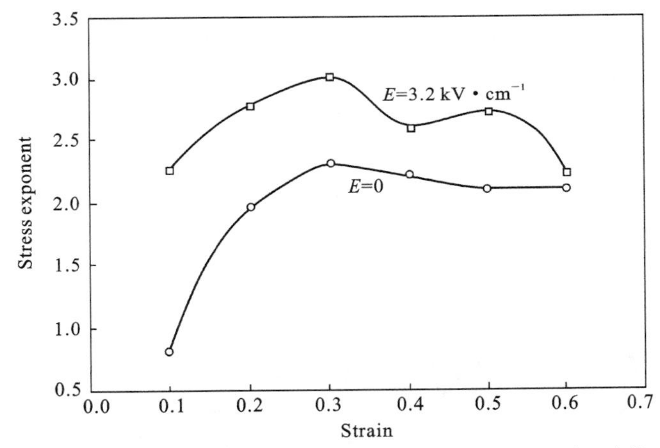

Figure 6.9 The n versus ε curve of negative electroplasticity.

6.3.4 Effect of electric field on diffusion dissolution zone

During superplastic deformation, there seems no difference between negatively charged specimen and positively charged specimen. It is shown by scanning electron microscope (SEM) and energy dispersive analysis of X-ray (EDAX) that no matter what the surface of specimen is charged, the amount of Zn and Al in the specimen has no change, and the surface of specimen has also no difference under the two circumstances. However, there exist certain differences between the surface of charged specimen and that of uncharged specimen. As shown in Figure 6.10, before superplastic deformation, the specimens are polished for viewing the change of surface. After superplastic deformation, large quantities of round voids occur on the surface of charged specimen, but for the uncharged specimen, the voids on the surface are less and smaller.

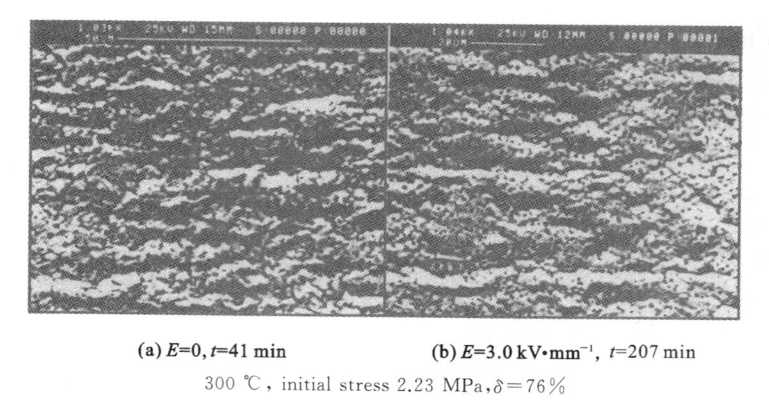

(a) E=0, t=41 min (b) E=3.0 kV·mm^{-1}, t=207 min

300 ℃, initial stress 2.23 MPa, δ=76%

Figure 6.10 Surface pattern of Zn-5Al alloy after superplastic deformation.

Figure 6.11 shows the TEM images of the specimen deformed superplastically with and without electric field for the rolled Zn-5Al alloy. Let the temperature of specimens reach 275 ℃ in 10 minutes, and then the specimens are deformed under the initial stress of 2.23 MPa. The specimen with electric field obtains an elongation of 343% in 76 minutes, and the specimen without electric field obtains an elongation of 670% in 26 minutes. As soon as the tensile test is finished, the specimen is taken out of the furnace and cooled in air, and then the specimens are kept at room temperature. Three months later, the specimens are viewed under TEM. It can be seen from Figure 6.11 that large quantities of precipitated particles occur in the specimens with electric field and seldom in the specimens without electric field. Obviously, the condition that the precipitating process can take place is solid solution treatment.

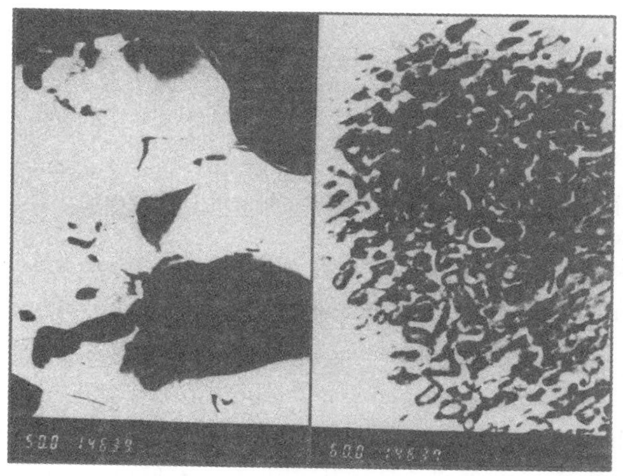

(a) E=0, δ=670%, t=26 min (b) E=3.0 kV·mm^{-1}, δ=343%, t=76 min

275 ℃, initial stress 2.23 MPa

Figure 6.11 Effect of electric field on solid solution precipitation of Zn-5Al alloy.

As can be seen, the main difference of Figures 6.11(a) and (b) lies in the strain rate. The former obtains the elongation of 670% at higher strain rate within 26 minutes; the latter obtains the elongation of 343% at lower strain rate within 76 minutes. For solid solution process, the longer time maybe be a favorable factor. However in 76 minutes, the electric field of 3.0 kV/mm and the leakage current of 1.85 $\times 10^{-2}$ A/m^2 are even more beneficial to the solid solution process.

The research about the electroplastic effect employed electric current and electric field has shown that the electric current and field can promote the diffusion of atoms in the specimen[58]. Therefore, the electric field advantages dissolution of Zn in α, and makes the thickness of diffusion dissolution zone increase, which is disadvantageous to the sliding of interface α/β. The result is that the strain rate of superplastic deformation is decreased. Meanwhile, thicker diffusion dissolution zone means that more Zn atoms dissolve in α phase. Afterwards, the Zn atoms precipitate from the diffusion dissolution zone, as shown in Figure 6.11(b).

6.4 Summary

During superplastic deformation in an electric field, Zn-5Al alloy exhibits a negative electroplastic effect that the elongation and the strain rate decrease and the stress of the deformation and the voids on the specimen's surface increase.

The most obvious effect occurs near the eutectoid temperature and the electric field can decrease the strain rate by an order of magnitude. The experimental results indicate that, during the initial deformation, the electric field makes many voids occur on the specimen's surface and somewhat influences the activation energy at high temperature.

The electric field and the leakage current help the atoms with their diffusion, and increase the thickness of the diffusion dissolution zone, which is disadvantageous to the sliding of the interface α/β. The result is that the strain rate of superplastic deformation is decreased.

Chapter 7

Effect of Rare Earth on Diffusion Dissolution Zone

Zn-5Al is a two-phase alloy, and mass ratio of α phase to β phase is 5 : 95. The mechanism of superplastic deformation for Zn-5Al alloy is the interface sliding controlled by the diffusion dissolution zone. All factors that can affect the diffusion and dissolution will affect the superplastic deformation process.

In this chapter, the effect of rare earth (RE) on the superplasticity in single-phase Al-4Zn-1Mg alloy is first introduced. Then the effect of rare earth on the superplasticity in Zn-5Al binary phase alloy is discussed.

7.1 Effect of RE on superplasticity in Al-4Zn-1Mg alloy

There were few reports about the application of rare earth elements in superplastic alloys, and in 1988 present author reported some experimental results that rare earth improves the superplasticity in Al-4Zn-1Mg alloys[64]. Single-phase Al-4Zn-1Mg mainly has reinforcing phase $MgZn_2$, and its mechanism of superplastic deformation is grain boundary sliding. For the comparison with Zn-5Al alloy, some results that rare earth affects superplasticity in Al-4Zn-1Mg alloy will be introduced and discussed below in a new perspective.

7.1.1 Amount of RE in Al-4Zn-1Mg alloy

Al-4Zn-1Mg alloy is prepared in a laboratory. Firstly, Al ingots of purity of 99.99 mass percent are melted. Secondly, Mn and Cr are joined in the way of intermediate alloy. Then, Zn and Mg are added at 725 ℃. Finally, rare earth metal is put into molten bath at 730 ℃. The composition of rare earth is 53.5% Ce, 30.8%La, 10.9%Nd, 3.9% Pr, 0.8% Sm and 0.1% Eu in mass percent. The alloys are cast at 720 ℃ and their compositions are shown in Table 7.1.

Table 7.1 Compositions of Al-4Zn-1Mg-RE alloys

Alloys	RE	Zn	Mg	Mn	Cr	Fe	Si	Al
Alloy 1	0.000	3.91	1.11	0.32	0.12	0.15	<0.2	Balance
Alloy 2	0.058	3.84	1.15	0.34	0.12	0.13	<0.2	Balance
Alloy 3	0.108	3.55	1.05	0.29	0.11	0.18	<0.2	Balance
Alloy 4	0.208	3.75	1.01	0.31	0.12	0.15	<0.2	Balance
Alloy 5	0.489	3.69	1.07	0.32	0.13	0.11	<0.2	Balance
Alloy 6	0.964	3.61	1.07	0.31	0.14	0.16	<0.2	Balance

7.1.2 Effect of RE on microstructure in Al-4Zn-1Mg alloy

The ingots of ϕ48 mm are homogenizingly treatmented by 465 ℃×19 h, and then a thermomechanical treatment is followed, which includes the deformation by 75% at 430 ℃, the solid solution by 460 ℃×2 h, the overaging by 350 ℃×24 h and the cold deformation by 80%, and finally the superplastic sheet of 2 mm-thickness is obtained.

The specimens for metallographic analysis are electrolyzedly polished with 10% perchloric acid and 90% alcohol, and then they are etched with a mixed solution of 0.5%HF, 1.5%HCl, 2.5%HNO_3 and 95.5%H_2O.

Rare earth can make cast structure be fine. It is known that rare earth in the Al-4Zn-1Mg alloy mainly distributes along grain boundaries according to the probe analysis. The solubility of rare earth in Al is so small that it does not excess 0.01% even at 500 ℃. Therefore, it is shown through X-ray diffraction that rare earth in Al-4Zn-1Mg exists in the form of compounds Al_2CeZn_2.

Rare earth can obviously reduce the grain size of recrystallized alloy, and the greater the amount of deformation of cold work, the finer the structure of the alloy. The recrystallized structures with rare earth and that without rare earth for Al-Zn-Mg alloys are shown in Figure 7.1.

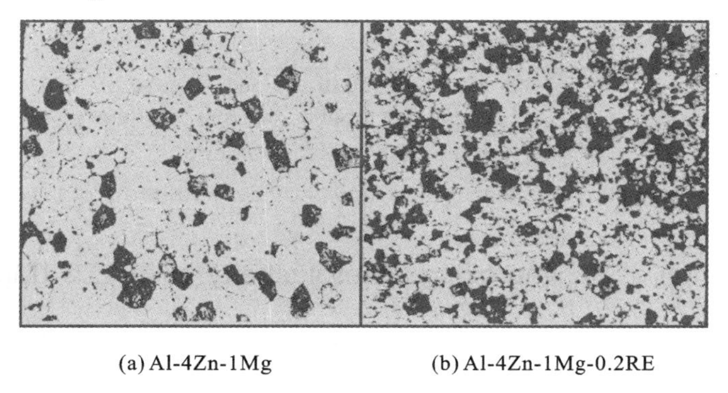

(a) Al-4Zn-1Mg (b) Al-4Zn-1Mg-0.2RE

Figure 7.1 Structures in Al-4Zn-1Mg alloys recrystallized by 540 ℃×1 h, 200×.

7.1.3 Effect of RE on superplasticity in Al-4Zn-1Mg alloy

Figure 7.2 is a plot of elongation versus deformation temperature for Al-4Zn-1Mg alloy at high strain rate of 1.10×10^{-2}/s, and it is shown that rare earth can obviously increase the elongation of superplastic deformation at the temperature of 580 ℃. The results at low strain rate of 3.60×10^{-4}/s are shown in Figure 7.3, which shows that the elongation of the specimen with 0.2 mass percent rare earth is larger than that with 1.0 mass percent rare earth. Hence rare earth has a creep-resisting ability during low strain rate deformation.

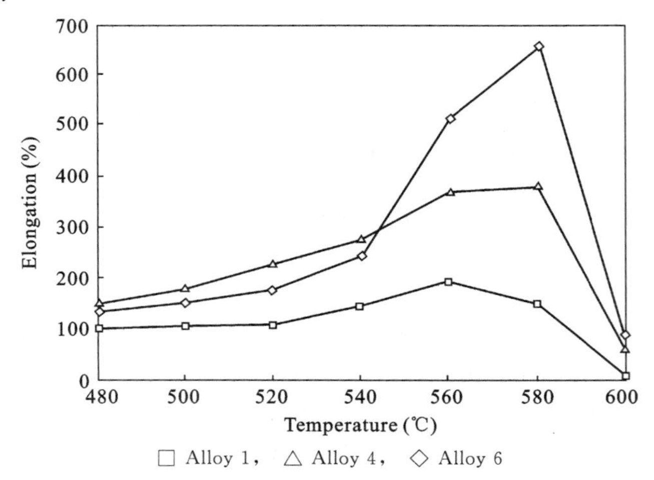

Figure 7.2 Elongation vs. temperature curves at high strain rate of 1.10×10^{-2}/s.

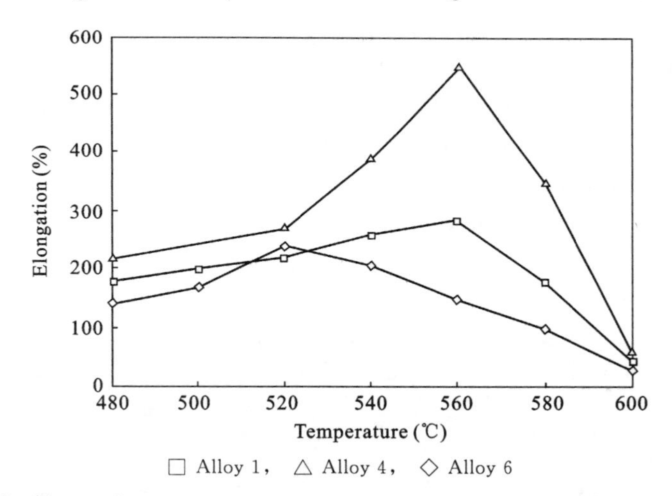

Figure 7.3 Elongation vs. temperature curves at low strain rate of 3.60×10^{-4}/s.

7.1.4 Explanation of RE improving superplasticity in Al-4Zn-1Mg alloy

The grain-size factor is an important condition for superplasticity, and it includes grain size, grain shape and the distribution of grain sizes. Figure 7.1 shows rare earth can obviously improve the grain size factor of the alloy.

The effects of rare earth on the textures in cold rolled specimen and superplastically deformed specimen are shown in Figures 7.4, 7.5 and 7.6. For cold rolled state, the orientation of grains in alloy 1 without rare earth is $(110)[1\bar{1}2]$ and that in alloys with rare earth of greater than 0.1% is (110)[001]. For face-centered cubic metals, $(110)[1\bar{1}0]$, (110)[001], $(112)[11\bar{1}]$ and $(110)[1\bar{1}2]$ are alined by the stability from low to high[65]. It is seen that the specimen without rare earth after being rolled at room temperature has a stable grain orientation $(110)[1\bar{1}2]$, and that with rare earth greater than 0.1% has a metastable grain orientation (110)[001].

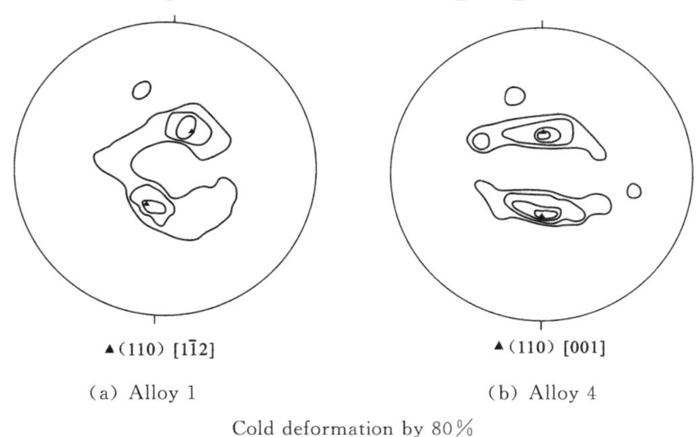

Figure 7.4 The partial (111) pole figure for rolled Al-4Zn-1Mg.

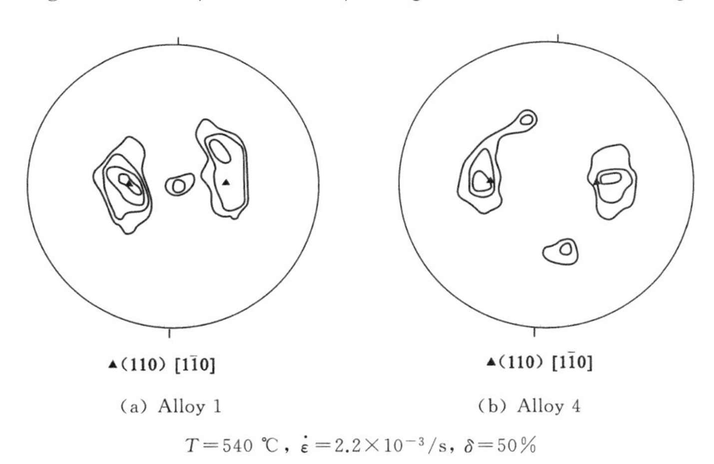

Figure 7.5 The partial (111) pole figure for superplastically deformed Al-4Zn-1Mg.

The rare earth in Al-4Zn-1Mg not only increases the grain boundaries in the alloy, but also affects the grain orientation in the alloy. The unstable grain orientation causes the grains to rotate, which is advantageous to grain boundaries sliding during the superplastic deformation.

7.2 Effect of RE on superplasticity in Zn-5Al alloy

The application of rare earth in Zn-5Al superplastic alloy was seldom reported, and present author reported the experimental results about rare earth improving the

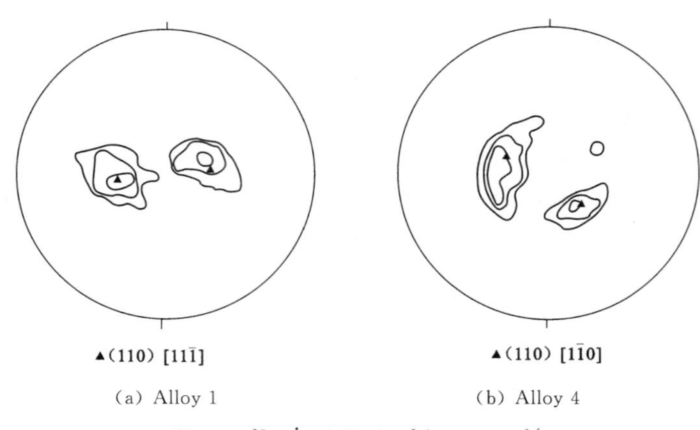

(a) Alloy 1 (b) Alloy 4

$T=540$ ℃, $\dot{\varepsilon}=2.2\times10^{-3}$/s, $\delta=100\%$

Figure 7.6 The partial (111) pole figure for superplastically deformed Al-4Zn-1Mg.

superplasticity in Zn-5Al alloy[66]. After that, Zhou Tairui[67] reported the application of rare earth in Zn-22Al alloy. It is said in this report that 0.05% rare earth in Zn-22Al can increase the strength, the hardness and the wear resistance of the alloy by 8%, 11% and 8% respectively. However about the superplasticity in this report, only strain rate sensitivity exponent m is concerned and there is little changes in m for the alloys with rare earth and without rare earth at the temperature of 250 ℃ and the strain rate of 5×10^{-2}/s.

7.2.1 Amount of RE in Zn-5Al alloy

Zn-5Al-RE alloys were prepared in a small factory. Firstly, Zn ingots were melted. Then, Al and Mg were put in. Finally, rare earth was joined in the way of Zn-Al-RE intermediate alloy. The ingots of 300 mm×400 mm×24 mm were obtained by direct chill casting method at between 450 ℃ and 500 ℃. The compositions of the alloys are shown in Table 7.2.

Table 7.2 Compositions of Zn-5Al-RE alloys

Alloys	RE	Al	Mg	Zn
Alloy 1	0.00	5.0	0.030	Balance
Alloy 2	0.05	4.8	0.027	Balance
Alloy 3	0.11	5.0	0.031	Balance
Alloy 4	0.16	5.4	0.029	Balance
Alloy 5	0.46	5.0	0.030	Balance
Alloy 6	0.61	5.3	0.032	Balance

7.2.2 Effect of RE on microstructure in Zn-5Al alloy

The cast plates of alloy are treated by homogenizing of 350 ℃×8 h, and then a hot rolled process is followed by the deformation of 73%. The homogenizing of 310 ℃×4 h and the deformation at lower temperature are followed again, and finally the sheets of thickness of 2 mm and 4 mm are obtained.

The solubility of Ce in Zn is too small to reach 0.01 mass percent even at 421 ℃. Rare earth in Zn-5Al exists in the form of the compounds with Al and Zn. Rare earth in Zn-5Al can obviously reduce the grain size of the alloy in cast state, but when rare earth exceeds 0.16%, larger compounds with rare earth will exist in the structure. These compounds are analyzed to be Al_2CeZn_2 and $CeZn_3$ through X-ray diffraction. No rare earth in eutectic structure can be found by X-ray energy spectrometer under SEM. The structures of the alloys recrystallized by 350 ℃×12 h are shown in Figure 7.7. It is seen that 0.05% rare earth in the alloy can obviously reduce the grain size of the recrystallized alloy and when the amount of rare earth exceeds 0.05%, the roles of rare earth have no increase.

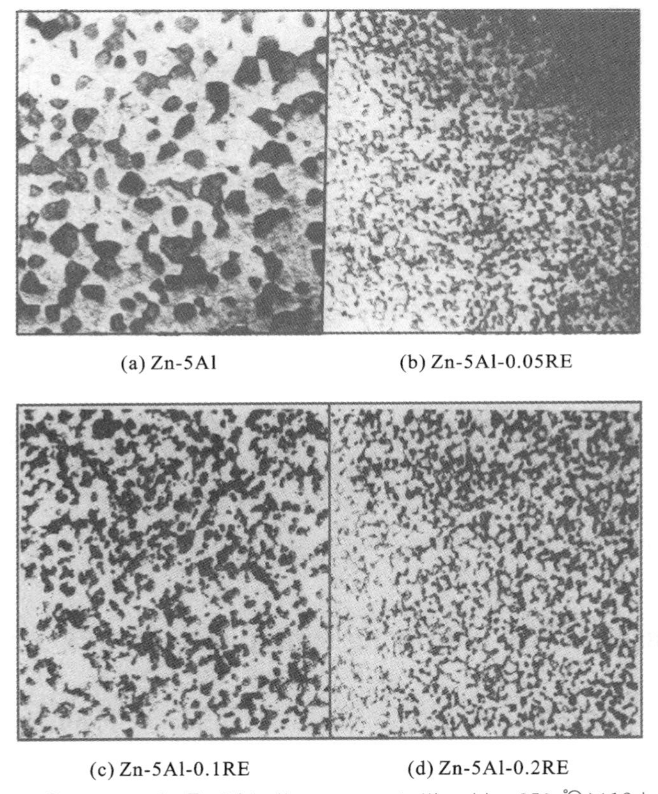

(a) Zn-5Al (b) Zn-5Al-0.05RE

(c) Zn-5Al-0.1RE (d) Zn-5Al-0.2RE

Figure 7.7 Structures in Zn-5Al alloys recrystallized by 350 ℃×12 h, 200×.

7.2.3 Effect of RE on superplasticity in Zn-5Al alloy

Figures 7.8, 7.9 and 7.10 are the plots of the elongation of superplastic deformation versus the amounts of rare earth for Zn-5Al alloys at 330 ℃, 350 ℃ and 370 ℃ respectively. Rare earth between 0.05 mass percent and 0.2 mass percent can obviously improve the superplasticity in Zn-5Al alloys above 350 ℃.

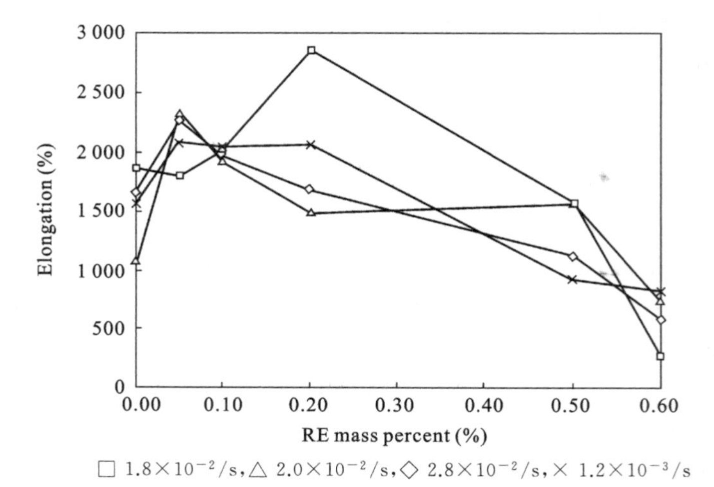

Figure 7.8 Results of superplastic deformation for Zn-5Al at 330 ℃.

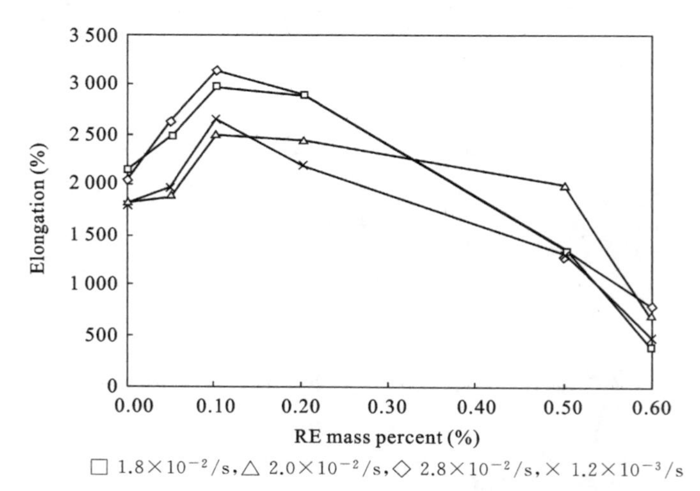

Figure 7.9 Results of superplastic deformation in Zn-5Al at 350 ℃.

7.3 Mechanism of RE improving superplasticity in Zn-5Al alloy

Al-4Zn-1Mg is a single-phase alloy, whose superplastic deformation mechanism is grain boundary sliding. Rare earth can increase grain boundaries in Al-4Zn-1Mg. Moreover, rare earth can raise melt temperature and recrystallized temperature of the alloy by 50 ℃. Therefore, rare earth has a stabilizing effect on the grain size of the alloys at the superplastic deformation temperature. The higher the deformation temperature, the larger the strain rate, and the more obvious effects of the rare earth improving the superplasticity of Al-4Zn-1Mg alloy.

7.3.1 Stabilization of RE on microstructure in Zn-5Al alloy

Zn-5Al is a two-phase alloy, and its mechanism for the superplastic deformation is

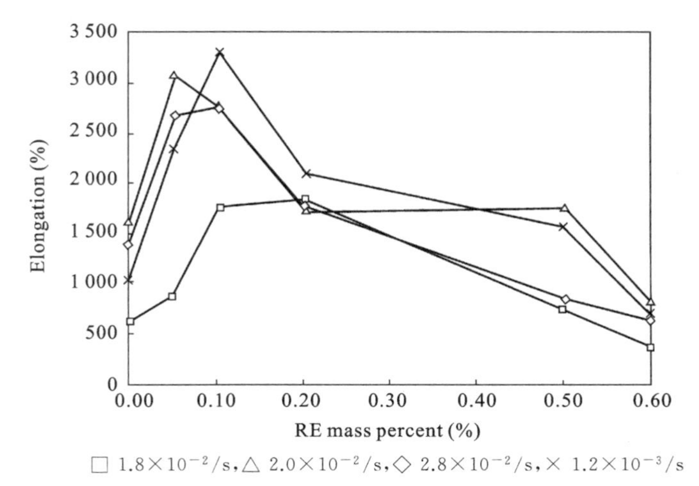

Figure 7.10 Results of superplastic deformation in Zn-5Al at 370 ℃.

the interface sliding controlled mainly by the diffusion dissolution zone. The mechanism of rare earth reducing the grain size in Zn-5Al is different from that in Al-4Zn-1Mg. It is shown by SEM that the localization of rare earth in Zn-5Al is also different from that in Al-4Zn-1Mg. Rare earth compounds are hardly seen in the grain boundaries or the interfaces for the former, and but rare earth compounds mainly are localized in the grain boundaries for the latter, as shown in Figure 7.11.

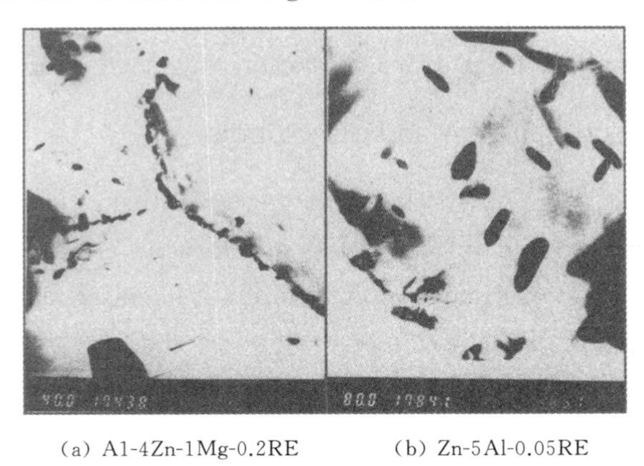

(a) Al-4Zn-1Mg-0.2RE (b) Zn-5Al-0.05RE

Figure 7.11 Configuration of rare earth compounds in alloys.

Rare earth compounds locate in the grain boundaries for Al-4Zn-1Mg alloy, which will obstructs the grains boundaries from migrating during recrystallization. Rare earth compound particles locate inside the grains for Zn-5Al alloy, and the mechanism that affects the recrystallization is different from that in Al-4Zn-1Mg. For Zn-5Al two-phase alloy, the recrystallization process will concern the diffusion and the dissolution of the atoms. Figure 7.12 is SEM photograph for Zn-5Al-0.05RE alloy which is deformed for 5 hours at 350 ℃ under the initial stress of 0.7 MPa. The face of the paper in the figure is the normal section of the specimen, and for SEM analysis the specimen was etched by

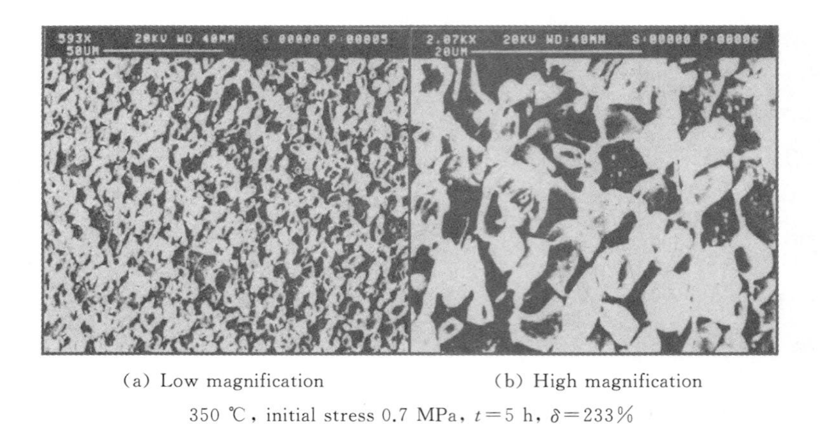

(a) Low magnification (b) High magnification

350 ℃, initial stress 0.7 MPa, $t=5$ h, $\delta=233\%$

Figure 7.12 Structures in Zn-5Al-0.05RE deformed for 5 h.

the method introduced in Section 3.2. Under the low magnification, it is shown with X-ray energy spectrometer that the proportion of Zn atoms to Al atoms is 88 against 12 on the surface of the etched specimen. For the protuberant white "grains", the proportion of Zn atoms to Al atoms is 74 against 26. The solubility of Zn in Al is 1 atom percent at room temperature, and it is not greater than 65 atom percent even at 350 ℃. Consequently, the protuberant white "grain", in fact, is a composite grain, which is an eutectoid structure that the bared parts are phase α and the covered parts are phase β. Obviously, rare earth can keep these composite grains from growing, which shows rare earth can obstruct Zn phase from diffusing and dissolving in Al phase.

7.3.2 Effect of RE on diffusion and dissolution

It is concluded in Chapter 4 that there is a diffusion dissolution zone in α/β interface, and the unsaturated diffusion dissolution zone is advantageous to interface sliding and saturated diffusion dissolution zone is disadvantageous to interface sliding. At 370 ℃, the elongation of the specimens (alloy 2 and alloy 3) with rare earth reaches 2500%, while that of the specimens without rare earth is 1000%. At this temperature, for alloy 1 without rare earth, due to great diffusion ability of Zn, the diffusion dissolution zone in interface α/β becomes saturated and thick, which is disadvantageous to interface sliding. For alloy 2 and alloy 3 with rare earth, rare earth keeps Zn from diffusing, and as a result the diffusion dissolution zone is unsaturated and thin, which is advantageous to interface sliding. Hence rare earth improves the superplasticity of alloys.

7.4 Summary

It is realized in Zn-5Al alloy by adding suitable amount of rare earth that the elongation during superplastic deformation is increased, and when the deformation temperature is over 350 ℃, rare earth can obviously enhance the superplastic effect in the alloy. At that temperature, rare earth can keep Zn phase from diffusing and dissolving in Al phase, and as a result the diffusion dissolution zone cannot become saturated and thick, which is advantageous to interface sliding. Consequently, the superplasticity in Zn-5Al alloy is improved obviously.

Appendix A

The Data of the Interaction Volume of Atoms

Atomic number is listed in the first column; Wigner-Seitz atomic radius is listed in the third column; closely-packed radius of atom is listed in the fourth column; atom interaction volume[8] is listed in the last column.

Unit: atomic radius in 10^{-10} m, interaction volume of atom in 10^{-30} m^3.

Table A.1 Interaction volume of atoms

Z	Name	r_{ws}	r_m	V_{act}
3	Li	1.73	1.56	5.85
4	Be	1.26	1.13	2.26
11	Na	2.11	1.91	10.16
12	Mg	1.77	1.60	6.09
13	Al	1.58	1.43	4.35
14	Si	1.69	1.32	10.46
19	K	2.62	2.38	18.75
20	Ca	2.28	1.98	17.13
21	Sc	1.81	1.64	6.49
22	Ti	1.61	1.46	4.56
23	V	1.49	1.35	3.55
24	Cr	1.42	1.28	3.21
25	Mn	1.43	1.26	3.89
26	Fe	1.41	1.27	3.21
27	Co	1.38	1.25	2.94
28	Ni	1.38	1.25	2.78
29	Cu	1.41	1.28	3.01
30	Zn	1.54	1.39	4.03
31	Ga	1.67	1.41	7.85
32	Ge	1.75	1.37	11.81

Continued

Z	Name	r_{ws}	r_m	V_{act}
33	As	1.73	1.39	10.50
37	Rb	2.81	2.55	23.37
38	Sr	2.37	2.15	14.33
39	Y	1.99	1.80	8.70
40	Zr	1.77	1.60	6.25
41	Nb	1.62	1.47	4.63
42	Mo	1.55	1.40	4.12
43	Tc	1.50	1.36	3.72
44	Ru	1.49	1.34	3.70
45	Rh	1.49	1.35	3.48
46	Pd	1.52	1.38	3.77
47	Ag	1.60	1.45	4.33
48	Cd	1.73	1.57	5.54
49	In	1.84	1.66	6.91
50	Sn	1.86	1.55	11.47
51	Sb	1.94	1.59	13.71
52	Te	2.01	1.73	12.35
55	Cs	3.03	2.73	31.01
56	Ba	2.49	2.24	17.68
57	La	2.07	1.88	9.58
58	Ce	2.02	1.82	9.11
59	Pr	2.02	1.83	8.89
60	Nd	2.01	1.82	8.96
62	Sm	1.99	1.80	8.70
63	Eu	2.26	1.80	23.61
64	Gd	1.99	1.80	8.63
65	Tb	1.97	1.78	8.43
66	Dy	1.96	1.77	8.37
67	Ho	1.95	1.77	7.95
68	Er	1.94	1.76	7.87
69	Tm	1.93	1.75	7.67
70	Yb	2.14	1.76	18.37
71	Lu	1.92	1.72	8.23
72	Hf	1.75	1.58	6.06
73	Ta	1.63	1.47	4.79
74	W	1.56	1.41	4.08

Continued

Z	Name	r_{ws}	r_m	V_{act}
75	Re	1.52	1.38	3.69
76	Os	1.50	1.35	3.69
77	Ir	1.50	1.36	3.64
78	Pt	1.53	1.39	3.86
79	Au	1.59	1.44	4.43
80	Hg	1.80	1.57	8.37
81	Tl	1.90	1.72	7.24
82	Pb	1.94	1.75	7.94
83	Bi	2.04	1.70	14.79
84	Po	2.08	1.76	14.85
90	Th	1.99	1.80	8.61

Appendix B

The Data of Binary Atomic Phase Diagram

The equilibrium electron densities between Li atom and X atom are listed in the last column in the table, for example, and first row is the data of Li itself. The first column is the atomic number of X atom; the second column is the name of X atom; the third column is the atomic radius of X atom; the fourth column is the electron density of X atom; the fifth column is the cohesive energy of X atom from reference [68], the sixth column is the equilibrium electron density.

Unit: atomic radius in 10^{-10} m, electron density in $10^{29}/m^3$, cohesive energy in kJ/mol.

Table B.1 Li-X

Z	Name	r	n	E	n_P
3	Li	1.7316	0.24	158	0.24
4	Be	1.256	1.44	320	1.04
11	Na	2.11	0.18	107	0.21
12	Mg	1.7706	0.57	145	0.39
13	Al	1.5825	1.07	327	0.80
14	Si	1.6864	0.81	446	0.66
19	K	2.6187	0.07	90.1	0.18
20	Ca	2.28	0.15	178	0.19
21	Sc	1.8132	0.69	37.6	0.32
22	Ti	1.6136	1.33	468	1.05
23	V	1.4901	2.04	512	1.61
24	Cr	1.4202	2.64	395	1.96
25	Mn	1.4308	2.61	282	1.76
26	Fe	1.4119	2.85	413	2.12
27	Co	1.3846	3.20	424	2.39
28	Ni	1.378	3.34	428	2.51

Continued

Z	Name	r	n	E	n_P
29	Cu	1.4119	3.03	336	2.13
30	Zn	1.5394	2.00	130	1.03
31	Ga	1.6723	1.33	271	0.93
32	Ge	1.7534	1.05	372	0.81
33	As	1.7316	1.14	285.3	0.82
37	Rb	2.8088	0.07	82.2	0.18
38	Sr	2.3728	0.17	166	0.20
39	Y	1.9924	0.56	422	0.47
40	Zr	1.7746	1.10	603	0.92
41	Nb	1.6237	1.79	730	1.52
42	Mo	1.5504	2.31	658	1.91
43	Tc	1.5043	2.72	661	2.24
44	Ru	1.4873	2.92	650	2.39
45	Rh	1.4873	2.95	554	2.35
46	Pd	1.5224	2.65	376	1.94
47	Ag	1.5982	2.09	284	1.42
48	Cd	1.7316	1.37	112	0.71
49	In	1.8394	1.00	243	0.70
50	Sn	1.8627	0.94	303	0.70
51	Sb	1.9393	0.75	265	0.56
52	Te	2.0105	0.61	215	0.45
55	Cs	3.0275	0.07	77.6	0.18
56	Ba	2.4912	0.14	183	0.19
57	La	2.0748	0.53	431	0.45
58	Ce	2.0168	0.63	417	0.52
59	Pr	2.0207	0.63	357	0.51
60	Nd	2.0138	0.65	328	0.51
62	Sm	1.9924	0.70	206	0.50
63	Eu	2.2551	0.32	179	0.28
64	Gd	1.991	0.71	400	0.58
65	Tb	1.9705	0.76	391	0.61
66	Dy	1.9612	0.79	294	0.60
67	Ho	1.9525	0.82	302	0.62
68	Er	1.9425	0.85	317	0.64
69	Tm	1.9301	0.89	233	0.62
70	Yb	2.1427	0.47	154	0.35

Continued

Z	Name	r	n	E	n_P
71	Lu	1.9177	0.93	428	0.74
72	Hf	1.7534	1.55	621	1.29
73	Ta	1.6288	2.33	782	1.98
74	W	1.5573	2.97	859	2.54
75	Re	1.5196	3.39	775	2.86
76	Os	1.4951	3.71	788	3.13
77	Ir	1.5015	3.66	670	3.00
78	Pt	1.5336	3.29	564	2.63
79	Au	1.5932	2.71	368	1.97
80	Hg	1.8037	1.39	65	0.57
81	Tl	1.8962	1.05	182	0.67
82	Pb	1.9359	0.94	196	0.63
83	Bi	2.0363	0.70	210	0.50
84	Po	2.0799	0.62	144	0.42
90	Th	1.9906	0.83	598	0.71

Table B.2 Be-X

Z	Name	r	n	E	n_P
4	Be	1.256	1.44	320	1.44
11	Na	2.11	0.18	107	1.12
12	Mg	1.7706	0.57	145	1.17
13	Al	1.5825	1.07	327	1.25
14	Si	1.6864	0.81	446	1.07
19	K	2.6187	0.07	90.1	1.14
20	Ca	2.28	0.15	178	0.98
21	Sc	1.8132	0.69	37.6	1.36
22	Ti	1.6136	1.33	468	1.37
23	V	1.4901	2.04	512	1.81
24	Cr	1.4202	2.64	395	2.10
25	Mn	1.4308	2.61	282	1.99
26	Fe	1.4119	2.85	413	2.23
27	Co	1.3846	3.20	424	2.44
28	Ni	1.378	3.34	428	2.53
29	Cu	1.4119	3.03	336	2.25
30	Zn	1.5394	2.00	130	1.60
31	Ga	1.6723	1.33	271	1.39

Continued

Z	Name	r	n	E	n_P
32	Ge	1.7534	1.05	372	1.23
33	As	1.7316	1.14	285.3	1.30
37	Rb	2.8088	0.07	82.2	1.16
38	Sr	2.3728	0.17	166	1.00
39	Y	1.9924	0.56	422	0.94
40	Zr	1.7746	1.10	603	1.22
41	Nb	1.6237	1.79	730	1.68
42	Mo	1.5504	2.31	658	2.02
43	Tc	1.5043	2.72	661	2.30
44	Ru	1.4873	2.92	650	2.43
45	Rh	1.4873	2.95	554	2.40
46	Pd	1.5224	2.65	376	2.09
47	Ag	1.5982	2.09	284	1.74
48	Cd	1.7316	1.37	112	1.42
49	In	1.8394	1.00	243	1.25
50	Sn	1.8627	0.94	303	1.19
51	Sb	1.9393	0.75	265	1.13
52	Te	2.0105	0.61	215	1.11
55	Cs	3.0275	0.07	77.6	1.17
56	Ba	2.4912	0.14	183	0.97
57	La	2.0748	0.53	431	0.92
58	Ce	2.0168	0.63	417	0.98
59	Pr	2.0207	0.63	357	1.01
60	Nd	2.0138	0.65	328	1.04
62	Sm	1.9924	0.70	206	1.15
63	Eu	2.2551	0.32	179	1.04
64	Gd	1.991	0.71	400	1.04
65	Tb	1.9705	0.76	391	1.07
66	Dy	1.9612	0.79	294	1.13
67	Ho	1.9525	0.82	302	1.14
68	Er	1.9425	0.85	317	1.14
69	Tm	1.9301	0.89	233	1.21
70	Yb	2.1427	0.47	154	1.12
71	Lu	1.9177	0.93	428	1.15
72	Hf	1.7534	1.55	621	1.51
73	Ta	1.6288	2.33	782	2.07

Continued

Z	Name	r	n	E	n_P
74	W	1.5573	2.97	859	2.55
75	Re	1.5196	3.39	775	2.82
76	Os	1.4951	3.71	788	3.06
77	Ir	1.5015	3.66	670	2.94
78	Pt	1.5336	3.29	564	2.62
79	Au	1.5932	2.71	368	2.12
80	Hg	1.8037	1.39	65	1.43
81	Tl	1.8962	1.05	182	1.30
82	Pb	1.9359	0.94	196	1.25
83	Bi	2.0363	0.70	210	1.15
84	Po	2.0799	0.62	144	1.18
90	Th	1.9906	0.83	598	1.04

Table B.3 Na-X

Z	Name	r	n	E	n_P
11	Na	2.11	0.18	107	0.18
12	Mg	1.7706	0.57	145	0.40
13	Al	1.5825	1.07	327	0.85
14	Si	1.6864	0.81	446	0.69
19	K	2.6187	0.07	90.1	0.13
20	Ca	2.28	0.15	178	0.17
21	Sc	1.8132	0.69	37.6	0.32
22	Ti	1.6136	1.33	468	1.11
23	V	1.4901	2.04	512	1.72
24	Cr	1.4202	2.64	395	2.12
25	Mn	1.4308	2.61	282	1.94
26	Fe	1.4119	2.85	413	2.30
27	Co	1.3846	3.20	424	2.59
28	Ni	1.378	3.34	428	2.71
29	Cu	1.4119	3.03	336	2.34
30	Zn	1.5394	2.00	130	1.18
31	Ga	1.6723	1.33	271	1.00
32	Ge	1.7534	1.05	372	0.85
33	As	1.7316	1.14	285.3	0.88
37	Rb	2.8088	0.07	82.2	0.13
38	Sr	2.3728	0.17	166	0.18

Continued

Z	Name	r	n	E	n_P
39	Y	1.9924	0.56	422	0.48
40	Zr	1.7746	1.10	603	0.96
41	Nb	1.6237	1.79	730	1.59
42	Mo	1.5504	2.31	658	2.01
43	Tc	1.5043	2.72	661	2.37
44	Ru	1.4873	2.92	650	2.53
45	Rh	1.4873	2.95	554	2.50
46	Pd	1.5224	2.65	376	2.10
47	Ag	1.5982	2.09	284	1.56
48	Cd	1.7316	1.37	112	0.79
49	In	1.8394	1.00	243	0.75
50	Sn	1.8627	0.94	303	0.74
51	Sb	1.9393	0.75	265	0.59
52	Te	2.0105	0.61	215	0.47
55	Cs	3.0275	0.07	77.6	0.14
56	Ba	2.4912	0.14	183	0.16
57	La	2.0748	0.53	431	0.46
58	Ce	2.0168	0.63	417	0.54
59	Pr	2.0207	0.63	357	0.53
60	Nd	2.0138	0.65	328	0.53
62	Sm	1.9924	0.70	206	0.52
63	Eu	2.2551	0.32	179	0.27
64	Gd	1.991	0.71	400	0.60
65	Tb	1.9705	0.76	391	0.64
66	Dy	1.9612	0.79	294	0.63
67	Ho	1.9525	0.82	302	0.65
68	Er	1.9425	0.85	317	0.68
69	Tm	1.9301	0.89	233	0.66
70	Yb	2.1427	0.47	154	0.35
71	Lu	1.9177	0.93	428	0.78
72	Hf	1.7534	1.55	621	1.35
73	Ta	1.6288	2.33	782	2.07
74	W	1.5573	2.97	859	2.66
75	Re	1.5196	3.39	775	3.00
76	Os	1.4951	3.71	788	3.29
77	Ir	1.5015	3.66	670	3.18

Continued

Z	Name	r	n	E	n_P
78	Pt	1.5336	3.29	564	2.80
79	Au	1.5932	2.71	368	2.14
80	Hg	1.8037	1.39	65	0.64
81	Tl	1.8962	1.05	182	0.73
82	Pb	1.9359	0.94	196	0.67
83	Bi	2.0363	0.70	210	0.53
84	Po	2.0799	0.62	144	0.43
90	Th	1.9906	0.83	598	0.73

Table B.4 Mg-X

Z	Name	r	n	E	n_P
12	Mg	1.7706	0.57	145	0.57
13	Al	1.5825	1.07	327	0.92
14	Si	1.6864	0.81	446	0.75
19	K	2.6187	0.07	90.1	0.38
20	Ca	2.28	0.15	178	0.34
21	Sc	1.8132	0.69	37.6	0.59
22	Ti	1.6136	1.33	468	1.15
23	V	1.4901	2.04	512	1.71
24	Cr	1.4202	2.64	395	2.09
25	Mn	1.4308	2.61	282	1.92
26	Fe	1.4119	2.85	413	2.25
27	Co	1.3846	3.20	424	2.53
28	Ni	1.378	3.34	428	2.64
29	Cu	1.4119	3.03	336	2.29
30	Zn	1.5394	2.00	130	1.25
31	Ga	1.6723	1.33	271	1.06
32	Ge	1.7534	1.05	372	0.91
33	As	1.7316	1.14	285.3	0.95
37	Rb	2.8088	0.07	82.2	0.39
38	Sr	2.3728	0.17	166	0.35
39	Y	1.9924	0.56	422	0.56
40	Zr	1.7746	1.10	603	1.00
41	Nb	1.6237	1.79	730	1.59
42	Mo	1.5504	2.31	658	1.99
43	Tc	1.5043	2.72	661	2.34

Continued

Z	Name	r	n	E	n_P
44	Ru	1.4873	2.92	650	2.49
45	Rh	1.4873	2.95	554	2.46
46	Pd	1.5224	2.65	376	2.07
47	Ag	1.5982	2.09	284	1.57
48	Cd	1.7316	1.37	112	0.92
49	In	1.8394	1.00	243	0.83
50	Sn	1.8627	0.94	303	0.82
51	Sb	1.9393	0.75	265	0.68
52	Te	2.0105	0.61	215	0.59
55	Cs	3.0275	0.07	77.6	0.39
56	Ba	2.4912	0.14	183	0.33
57	La	2.0748	0.53	431	0.54
58	Ce	2.0168	0.63	417	0.61
59	Pr	2.0207	0.63	357	0.61
60	Nd	2.0138	0.65	328	0.62
62	Sm	1.9924	0.70	206	0.65
63	Eu	2.2551	0.32	179	0.43
64	Gd	1.991	0.71	400	0.67
65	Tb	1.9705	0.76	391	0.71
66	Dy	1.9612	0.79	294	0.72
67	Ho	1.9525	0.82	302	0.74
68	Er	1.9425	0.85	317	0.76
69	Tm	1.9301	0.89	233	0.76
70	Yb	2.1427	0.47	154	0.52
71	Lu	1.9177	0.93	428	0.84
72	Hf	1.7534	1.55	621	1.37
73	Ta	1.6288	2.33	782	2.05
74	W	1.5573	2.97	859	2.62
75	Re	1.5196	3.39	775	2.95
76	Os	1.4951	3.71	788	3.22
77	Ir	1.5015	3.66	670	3.11
78	Pt	1.5336	3.29	564	2.74
79	Au	1.5932	2.71	368	2.11
80	Hg	1.8037	1.39	65	0.82
81	Tl	1.8962	1.05	182	0.84
82	Pb	1.9359	0.94	196	0.78

Continued

Z	Name	r	n	E	n_P
83	Bi	2.0363	0.70	210	0.65
84	Po	2.0799	0.62	144	0.59
90	Th	1.9906	0.83	598	0.78

Table B.5 Al-X

Z	Name	r	n	E	n_P
13	Al	1.5825	1.07	327	1.07
14	Si	1.6864	0.81	446	0.92
19	K	2.6187	0.07	90.1	0.86
20	Ca	2.28	0.15	178	0.75
21	Sc	1.8132	0.69	37.6	1.03
22	Ti	1.6136	1.33	468	1.22
23	V	1.4901	2.04	512	1.66
24	Cr	1.4202	2.64	395	1.93
25	Mn	1.4308	2.61	282	1.79
26	Fe	1.4119	2.85	413	2.06
27	Co	1.3846	3.20	424	2.27
28	Ni	1.378	3.34	428	2.36
29	Cu	1.4119	3.03	336	2.06
30	Zn	1.5394	2.00	130	1.34
31	Ga	1.6723	1.33	271	1.19
32	Ge	1.7534	1.05	372	1.06
33	As	1.7316	1.14	285.3	1.10
37	Rb	2.8088	0.07	82.2	0.87
38	Sr	2.3728	0.17	166	0.77
39	Y	1.9924	0.56	422	0.78
40	Zr	1.7746	1.10	603	1.09
41	Nb	1.6237	1.79	730	1.57
42	Mo	1.5504	2.31	658	1.90
43	Tc	1.5043	2.72	661	2.18
44	Ru	1.4873	2.92	650	2.30
45	Rh	1.4873	2.95	554	2.26
46	Pd	1.5224	2.65	376	1.92
47	Ag	1.5982	2.09	284	1.54
48	Cd	1.7316	1.37	112	1.15
49	In	1.8394	1.00	243	1.04

Continued

Z	Name	r	n	E	n_P
50	Sn	1.8627	0.94	303	1.01
51	Sb	1.9393	0.75	265	0.93
52	Te	2.0105	0.61	215	0.89
55	Cs	3.0275	0.07	77.6	0.88
56	Ba	2.4912	0.14	183	0.74
57	La	2.0748	0.53	431	0.76
58	Ce	2.0168	0.63	417	0.83
59	Pr	2.0207	0.63	357	0.84
60	Nd	2.0138	0.65	328	0.86
62	Sm	1.9924	0.70	206	0.93
63	Eu	2.2551	0.32	179	0.81
64	Gd	1.991	0.71	400	0.88
65	Tb	1.9705	0.76	391	0.91
66	Dy	1.9612	0.79	294	0.94
67	Ho	1.9525	0.82	302	0.95
68	Er	1.9425	0.85	317	0.96
69	Tm	1.9301	0.89	233	1.00
70	Yb	2.1427	0.47	154	0.88
71	Lu	1.9177	0.93	428	0.99
72	Hf	1.7534	1.55	621	1.39
73	Ta	1.6288	2.33	782	1.96
74	W	1.5573	2.97	859	2.45
75	Re	1.5196	3.39	775	2.71
76	Os	1.4951	3.71	788	2.94
77	Ir	1.5015	3.66	670	2.81
78	Pt	1.5336	3.29	564	2.48
79	Au	1.5932	2.71	368	1.94
80	Hg	1.8037	1.39	65	1.13
81	Tl	1.8962	1.05	182	1.07
82	Pb	1.9359	0.94	196	1.02
83	Bi	2.0363	0.70	210	0.93
84	Po	2.0799	0.62	144	0.94
90	Th	1.9906	0.83	598	0.92

Table B.6 Si-X

Z	Name	r	n	E	n_P
14	Si	1.6864	0.81	446	0.81

Continued

Z	Name	r	n	E	n_P
19	K	2.6187	0.07	90.1	0.68
20	Ca	2.28	0.15	178	0.62
21	Sc	1.8132	0.69	37.6	0.80
22	Ti	1.6136	1.33	468	1.07
23	V	1.4901	2.04	512	1.46
24	Cr	1.4202	2.64	395	1.67
25	Mn	1.4308	2.61	282	1.51
26	Fe	1.4119	2.85	413	1.79
27	Co	1.3846	3.20	424	1.97
28	Ni	1.378	3.34	428	2.05
29	Cu	1.4119	3.03	336	1.76
30	Zn	1.5394	2.00	130	1.08
31	Ga	1.6723	1.33	271	1.00
32	Ge	1.7534	1.05	372	0.92
33	As	1.7316	1.14	285.3	0.94
37	Rb	2.8088	0.07	82.2	0.69
38	Sr	2.3728	0.17	166	0.64
39	Y	1.9924	0.56	422	0.69
40	Zr	1.7746	1.10	603	0.98
41	Nb	1.6237	1.79	730	1.42
42	Mo	1.5504	2.31	658	1.70
43	Tc	1.5043	2.72	661	1.95
44	Ru	1.4873	2.92	650	2.06
45	Rh	1.4873	2.95	554	2.00
46	Pd	1.5224	2.65	376	1.65
47	Ag	1.5982	2.09	284	1.31
48	Cd	1.7316	1.37	112	0.92
49	In	1.8394	1.00	243	0.87
50	Sn	1.8627	0.94	303	0.86
51	Sb	1.9393	0.75	265	0.79
52	Te	2.0105	0.61	215	0.74
55	Cs	3.0275	0.07	77.6	0.70
56	Ba	2.4912	0.14	183	0.62
57	La	2.0748	0.53	431	0.67
58	Ce	2.0168	0.63	417	0.72
59	Pr	2.0207	0.63	357	0.73

Continued

Z	Name	r	n	E	n_P
60	Nd	2.0138	0.65	328	0.74
62	Sm	1.9924	0.70	206	0.77
63	Eu	2.2551	0.32	179	0.67
64	Gd	1.991	0.71	400	0.76
65	Tb	1.9705	0.76	391	0.79
66	Dy	1.9612	0.79	294	0.80
67	Ho	1.9525	0.82	302	0.81
68	Er	1.9425	0.85	317	0.83
69	Tm	1.9301	0.89	233	0.84
70	Yb	2.1427	0.47	154	0.72
71	Lu	1.9177	0.93	428	0.87
72	Hf	1.7534	1.55	621	1.24
73	Ta	1.6288	2.33	782	1.78
74	W	1.5573	2.97	859	2.23
75	Re	1.5196	3.39	775	2.45
76	Os	1.4951	3.71	788	2.66
77	Ir	1.5015	3.66	670	2.52
78	Pt	1.5336	3.29	564	2.20
79	Au	1.5932	2.71	368	1.67
80	Hg	1.8037	1.39	65	0.88
81	Tl	1.8962	1.05	182	0.88
82	Pb	1.9359	0.94	196	0.85
83	Bi	2.0363	0.70	210	0.77
84	Po	2.0799	0.62	144	0.76
90	Th	1.9906	0.83	598	0.82

Table B.7 K-X

Z	Name	r	n	E	n_P
19	K	2.6187	0.07	90.1	0.07
20	Ca	2.28	0.15	178	0.13
21	Sc	1.8132	0.69	37.6	0.57
22	Ti	1.6136	1.33	468	1.12
23	V	1.4901	2.04	512	1.74
24	Cr	1.4202	2.64	395	2.17
25	Mn	1.4308	2.61	282	1.99
26	Fe	1.4119	2.85	413	2.35

Continued

Z	Name	r	n	E	n_P
27	Co	1.3846	3.20	424	2.65
28	Ni	1.378	3.34	428	2.77
29	Cu	1.4119	3.03	336	2.40
30	Zn	1.5394	2.00	130	1.21
31	Ga	1.6723	1.33	271	1.01
32	Ge	1.7534	1.05	372	0.86
33	As	1.7316	1.14	285.3	0.88
37	Rb	2.8088	0.00	82.2	0.07
38	Sr	2.3728	0.17	166	0.14
39	Y	1.9924	0.56	422	0.47
40	Zr	1.7746	1.10	603	0.97
41	Nb	1.6237	1.79	730	1.60
42	Mo	1.5504	2.31	658	2.04
43	Tc	1.5043	2.72	661	2.40
44	Ru	1.4873	2.92	650	2.57
45	Rh	1.4873	2.95	554	2.55
46	Pd	1.5224	2.65	376	2.15
47	Ag	1.5982	2.09	284	1.60
48	Cd	1.7316	1.37	112	0.79
49	In	1.8394	1.00	243	0.74
50	Sn	1.8627	0.94	303	0.74
51	Sb	1.9393	0.75	265	0.58
52	Te	2.0105	0.61	215	0.45
55	Cs	3.0275	0.00	77.6	0.07
56	Ba	2.4912	0.14	183	0.12
57	La	2.0748	0.53	431	0.45
58	Ce	2.0168	0.63	417	0.53
59	Pr	2.0207	0.63	357	0.52
60	Nd	2.0138	0.65	328	0.52
62	Sm	1.9924	0.70	206	0.51
63	Eu	2.2551	0.32	179	0.24
64	Gd	1.991	0.71	400	0.60
65	Tb	1.9705	0.76	391	0.63
66	Dy	1.9612	0.79	294	0.62
67	Ho	1.9525	0.82	302	0.65
68	Er	1.9425	0.85	317	0.68

Continued

Z	Name	r	n	E	n_P
69	Tm	1.9301	0.89	233	0.66
70	Yb	2.1427	0.47	154	0.33
71	Lu	1.9177	0.93	428	0.78
72	Hf	1.7534	1.55	621	1.37
73	Ta	1.6288	2.33	782	2.09
74	W	1.5573	2.97	859	2.69
75	Re	1.5196	3.39	775	3.05
76	Os	1.4951	3.71	788	3.34
77	Ir	1.5015	3.66	670	3.23
78	Pt	1.5336	3.29	564	2.85
79	Au	1.5932	2.71	368	2.19
80	Hg	1.8037	1.39	65	0.62
81	Tl	1.8962	1.05	182	0.73
82	Pb	1.9359	0.94	196	0.67
83	Bi	2.0363	0.70	210	0.51
84	Po	2.0799	0.62	144	0.41
90	Th	1.9906	0.83	598	0.73

Table B.8 Ca-X

Z	Name	r	n	E	n_P
20	Ca	2.28	0.15	178	0.15
21	Sc	1.8132	0.69	37.6	0.25
22	Ti	1.6136	1.33	468	1.00
23	V	1.4901	2.04	512	1.55
24	Cr	1.4202	2.64	395	1.87
25	Mn	1.4308	2.61	282	1.66
26	Fe	1.4119	2.85	413	2.04
27	Co	1.3846	3.20	424	2.30
28	Ni	1.378	3.34	428	2.41
29	Cu	1.4119	3.03	336	2.03
30	Zn	1.5394	2.00	130	0.94
31	Ga	1.6723	1.33	271	0.86
32	Ge	1.7534	1.05	372	0.76
33	As	1.7316	1.14	285.3	0.76
37	Rb	2.8088	0.00	82.2	0.11
38	Sr	2.3728	0.17	166	0.16

Continued

Z	Name	r	n	E	n_P
39	Y	1.9924	0.56	422	0.44
40	Zr	1.7746	1.10	603	0.88
41	Nb	1.6237	1.79	730	1.47
42	Mo	1.5504	2.31	658	1.85
43	Tc	1.5043	2.72	661	2.18
44	Ru	1.4873	2.92	650	2.33
45	Rh	1.4873	2.95	554	2.27
46	Pd	1.5224	2.65	376	1.85
47	Ag	1.5982	2.09	284	1.34
48	Cd	1.7316	1.37	112	0.63
49	In	1.8394	1.00	243	0.64
50	Sn	1.8627	0.94	303	0.65
51	Sb	1.9393	0.75	265	0.51
52	Te	2.0105	0.61	215	0.40
55	Cs	3.0275	0.00	77.6	0.11
56	Ba	2.4912	0.14	183	0.15
57	La	2.0748	0.53	431	0.42
58	Ce	2.0168	0.63	417	0.49
59	Pr	2.0207	0.63	357	0.47
60	Nd	2.0138	0.65	328	0.47
62	Sm	1.9924	0.70	206	0.45
63	Eu	2.2551	0.32	179	0.24
64	Gd	1.991	0.71	400	0.54
65	Tb	1.9705	0.76	391	0.57
66	Dy	1.9612	0.79	294	0.55
67	Ho	1.9525	0.82	302	0.57
68	Er	1.9425	0.85	317	0.60
69	Tm	1.9301	0.89	233	0.57
70	Yb	2.1427	0.47	154	0.30
71	Lu	1.9177	0.93	428	0.70
72	Hf	1.7534	1.55	621	1.24
73	Ta	1.6288	2.33	782	1.93
74	W	1.5573	2.97	859	2.49
75	Re	1.5196	3.39	775	2.79
76	Os	1.4951	3.71	788	3.06
77	Ir	1.5015	3.66	670	2.92

Continued

Z	Name	r	n	E	n_P
78	Pt	1.5336	3.29	564	2.54
79	Au	1.5932	2.71	368	1.88
80	Hg	1.8037	1.39	65	0.49
81	Tl	1.8962	1.05	182	0.61
82	Pb	1.9359	0.94	196	0.57
83	Bi	2.0363	0.70	210	0.45
84	Po	2.0799	0.62	144	0.36
90	Th	1.9906	0.83	598	0.68

Table B.9 Sc-X

Z	Name	r	n	E	n_P
21	Sc	1.8132	0.69	37.6	0.69
22	Ti	1.6136	1.33	468	1.04
23	V	1.4901	2.04	512	1.47
24	Cr	1.4202	2.64	395	1.69
25	Mn	1.4308	2.61	282	1.51
26	Fe	1.4119	2.85	413	1.82
27	Co	1.3846	3.20	424	2.02
28	Ni	1.378	3.34	428	2.10
29	Cu	1.4119	3.03	336	1.79
30	Zn	1.5394	2.00	130	1.03
31	Ga	1.6723	1.33	271	0.96
32	Ge	1.7534	1.05	372	0.87
33	As	1.7316	1.14	285.3	0.89
37	Rb	2.8088	0.00	82.2	0.57
38	Sr	2.3728	0.17	166	0.53
39	Y	1.9924	0.56	422	0.62
40	Zr	1.7746	1.10	603	0.94
41	Nb	1.6237	1.79	730	1.42
42	Mo	1.5504	2.31	658	1.72
43	Tc	1.5043	2.72	661	1.99
44	Ru	1.4873	2.92	650	2.10
45	Rh	1.4873	2.95	554	2.04
46	Pd	1.5224	2.65	376	1.67
47	Ag	1.5982	2.09	284	1.29
48	Cd	1.7316	1.37	112	0.85

Continued

Z	Name	r	n	E	n_P
49	In	1.8394	1.00	243	0.81
50	Sn	1.8627	0.94	303	0.80
51	Sb	1.9393	0.75	265	0.72
52	Te	2.0105	0.61	215	0.66
55	Cs	3.0275	0.00	77.6	0.57
56	Ba	2.4912	0.14	183	0.51
57	La	2.0748	0.53	431	0.60
58	Ce	2.0168	0.63	417	0.66
59	Pr	2.0207	0.63	357	0.66
60	Nd	2.0138	0.65	328	0.67
62	Sm	1.9924	0.70	206	0.70
63	Eu	2.2551	0.32	179	0.57
64	Gd	1.991	0.71	400	0.70
65	Tb	1.9705	0.76	391	0.73
66	Dy	1.9612	0.79	294	0.74
67	Ho	1.9525	0.82	302	0.75
68	Er	1.9425	0.85	317	0.76
69	Tm	1.9301	0.89	233	0.77
70	Yb	2.1427	0.47	154	0.63
71	Lu	1.9177	0.93	428	0.82
72	Hf	1.7534	1.55	621	1.23
73	Ta	1.6288	2.33	782	1.80
74	W	1.5573	2.97	859	2.28
75	Re	1.5196	3.39	775	2.51
76	Os	1.4951	3.71	788	2.74
77	Ir	1.5015	3.66	670	2.59
78	Pt	1.5336	3.29	564	2.25
79	Au	1.5932	2.71	368	1.69
80	Hg	1.8037	1.39	65	0.80
81	Tl	1.8962	1.05	182	0.81
82	Pb	1.9359	0.94	196	0.78
83	Bi	2.0363	0.70	210	0.70
84	Po	2.0799	0.62	144	0.67
90	Th	1.9906	0.83	598	0.78

Table B.10 Ti-X

Z	Name	r	n	E	n_P
22	Ti	1.6136	1.33	468	1.33
23	V	1.4901	2.04	512	1.70
24	Cr	1.4202	2.64	395	1.93
25	Mn	1.4308	2.61	282	1.81
26	Fe	1.4119	2.85	413	2.04
27	Co	1.3846	3.20	424	2.22
28	Ni	1.378	3.34	428	2.29
29	Cu	1.4119	3.03	336	2.04
30	Zn	1.5394	2.00	130	1.47
31	Ga	1.6723	1.33	271	1.33
32	Ge	1.7534	1.05	372	1.20
33	As	1.7316	1.14	285.3	1.26
37	Rb	2.8088	0.00	82.2	1.13
38	Sr	2.3728	0.17	166	1.02
39	Y	1.9924	0.56	422	0.96
40	Zr	1.7746	1.10	603	1.20
41	Nb	1.6237	1.79	730	1.61
42	Mo	1.5504	2.31	658	1.90
43	Tc	1.5043	2.72	661	2.14
44	Ru	1.4873	2.92	650	2.25
45	Rh	1.4873	2.95	554	2.21
46	Pd	1.5224	2.65	376	1.92
47	Ag	1.5982	2.09	284	1.61
48	Cd	1.7316	1.37	112	1.34
49	In	1.8394	1.00	243	1.21
50	Sn	1.8627	0.94	303	1.17
51	Sb	1.9393	0.75	265	1.12
52	Te	2.0105	0.61	215	1.10
55	Cs	3.0275	0.00	77.6	1.14
56	Ba	2.4912	0.14	183	0.99
57	La	2.0748	0.53	431	0.94
58	Ce	2.0168	0.63	417	1.00
59	Pr	2.0207	0.63	357	1.02
60	Nd	2.0138	0.65	328	1.05
62	Sm	1.9924	0.70	206	1.14
63	Eu	2.2551	0.32	179	1.05

Continued

Z	Name	r	n	E	n_P
64	Gd	1.991	0.71	400	1.04
65	Tb	1.9705	0.76	391	1.07
66	Dy	1.9612	0.79	294	1.12
67	Ho	1.9525	0.82	302	1.13
68	Er	1.9425	0.85	317	1.13
69	Tm	1.9301	0.89	233	1.18
70	Yb	2.1427	0.47	154	1.12
71	Lu	1.9177	0.93	428	1.14
72	Hf	1.7534	1.55	621	1.46
73	Ta	1.6288	2.33	782	1.95
74	W	1.5573	2.97	859	2.39
75	Re	1.5196	3.39	775	2.62
76	Os	1.4951	3.71	788	2.82
77	Ir	1.5015	3.66	670	2.70
78	Pt	1.5336	3.29	564	2.40
79	Au	1.5932	2.71	368	1.94
80	Hg	1.8037	1.39	65	1.33
81	Tl	1.8962	1.05	182	1.25
82	Pb	1.9359	0.94	196	1.21
83	Bi	2.0363	0.70	210	1.13
84	Po	2.0799	0.62	144	1.16
90	Th	1.9906	0.83	598	1.05

Table B.11 V-X

Z	Name	r	n	E	n_P
23	V	1.4901	2.04	512	2.04
24	Cr	1.4202	2.64	395	2.30
25	Mn	1.4308	2.61	282	2.24
26	Fe	1.4119	2.85	413	2.40
27	Co	1.3846	3.20	424	2.56
28	Ni	1.378	3.34	428	2.63
29	Cu	1.4119	3.03	336	2.43
30	Zn	1.5394	2.00	130	2.03
31	Ga	1.6723	1.33	271	1.79
32	Ge	1.7534	1.05	372	1.62
33	As	1.7316	1.14	285.3	1.71

Continued

Z	Name	r	n	E	n_P
37	Rb	2.8088	0.00	82.2	1.75
38	Sr	2.3728	0.17	166	1.58
39	Y	1.9924	0.56	422	1.37
40	Zr	1.7746	1.10	603	1.53
41	Nb	1.6237	1.79	730	1.89
42	Mo	1.5504	2.31	658	2.19
43	Tc	1.5043	2.72	661	2.42
44	Ru	1.4873	2.92	650	2.53
45	Rh	1.4873	2.95	554	2.51
46	Pd	1.5224	2.65	376	2.30
47	Ag	1.5982	2.09	284	2.05
48	Cd	1.7316	1.37	112	1.92
49	In	1.8394	1.00	243	1.70
50	Sn	1.8627	0.94	303	1.63
51	Sb	1.9393	0.75	265	1.60
52	Te	2.0105	0.61	215	1.61
55	Cs	3.0275	0.00	77.6	1.77
56	Ba	2.4912	0.14	183	1.54
57	La	2.0748	0.53	431	1.35
58	Ce	2.0168	0.63	417	1.41
59	Pr	2.0207	0.63	357	1.46
60	Nd	2.0138	0.65	328	1.49
62	Sm	1.9924	0.70	206	1.65
63	Eu	2.2551	0.32	179	1.59
64	Gd	1.991	0.71	400	1.46
65	Tb	1.9705	0.76	391	1.49
66	Dy	1.9612	0.79	294	1.58
67	Ho	1.9525	0.82	302	1.58
68	Er	1.9425	0.85	317	1.58
69	Tm	1.9301	0.89	233	1.68
70	Yb	2.1427	0.47	154	1.68
71	Lu	1.9177	0.93	428	1.53
72	Hf	1.7534	1.55	621	1.77
73	Ta	1.6288	2.33	782	2.21
74	W	1.5573	2.97	859	2.62
75	Re	1.5196	3.39	775	2.85

Continued

Z	Name	r	n	E	n_P
76	Os	1.4951	3.71	788	3.05
77	Ir	1.5015	3.66	670	2.95
78	Pt	1.5336	3.29	564	2.70
79	Au	1.5932	2.71	368	2.32
80	Hg	1.8037	1.39	65	1.96
81	Tl	1.8962	1.05	182	1.78
82	Pb	1.9359	0.94	196	1.73
83	Bi	2.0363	0.70	210	1.65
84	Po	2.0799	0.62	144	1.73
90	Th	1.9906	0.83	598	1.39

Table B.12　Cr-X

Z	Name	r	n	E	n_P
24	Cr	1.4202	2.64	395	2.64
25	Mn	1.4308	2.61	282	2.63
26	Fe	1.4119	2.85	413	2.75
27	Co	1.3846	3.20	424	2.93
28	Ni	1.378	3.34	428	3.01
29	Cu	1.4119	3.03	336	2.82
30	Zn	1.5394	2.00	130	2.49
31	Ga	1.6723	1.33	271	2.11
32	Ge	1.7534	1.05	372	1.87
33	As	1.7316	1.14	285.3	2.01
37	Rb	2.8088	0.00	82.2	2.19
38	Sr	2.3728	0.17	166	1.91
39	Y	1.9924	0.56	422	1.57
40	Zr	1.7746	1.10	603	1.71
41	Nb	1.6237	1.79	730	2.09
42	Mo	1.5504	2.31	658	2.43
43	Tc	1.5043	2.72	661	2.69
44	Ru	1.4873	2.92	650	2.82
45	Rh	1.4873	2.95	554	2.82
46	Pd	1.5224	2.65	376	2.65
47	Ag	1.5982	2.09	284	2.41
48	Cd	1.7316	1.37	112	2.36
49	In	1.8394	1.00	243	2.02

Continued

Z	Name	r	n	E	n_P
50	Sn	1.8627	0.94	303	1.90
51	Sb	1.9393	0.75	265	1.88
52	Te	2.0105	0.61	215	1.93
55	Cs	3.0275	0.00	77.6	2.21
56	Ba	2.4912	0.14	183	1.85
57	La	2.0748	0.53	431	1.54
58	Ce	2.0168	0.63	417	1.61
59	Pr	2.0207	0.63	357	1.69
60	Nd	2.0138	0.65	328	1.74
62	Sm	1.9924	0.70	206	1.98
63	Eu	2.2551	0.32	179	1.92
64	Gd	1.991	0.71	400	1.67
65	Tb	1.9705	0.76	391	1.71
66	Dy	1.9612	0.79	294	1.85
67	Ho	1.9525	0.82	302	1.85
68	Er	1.9425	0.85	317	1.84
69	Tm	1.9301	0.89	233	1.99
70	Yb	2.1427	0.47	154	2.04
71	Lu	1.9177	0.93	428	1.75
72	Hf	1.7534	1.55	621	1.98
73	Ta	1.6288	2.33	782	2.43
74	W	1.5573	2.97	859	2.87
75	Re	1.5196	3.39	775	3.14
76	Os	1.4951	3.71	788	3.36
77	Ir	1.5015	3.66	670	3.28
78	Pt	1.5336	3.29	564	3.03
79	Au	1.5932	2.71	368	2.68
80	Hg	1.8037	1.39	65	2.47
81	Tl	1.8962	1.05	182	2.14
82	Pb	1.9359	0.94	196	2.08
83	Bi	2.0363	0.70	210	1.97
84	Po	2.0799	0.62	144	2.10
90	Th	1.9906	0.83	598	1.55

Table B.13 Mn-X

Z	Name	r	n	E	n_P
25	Mn	1.4308	2.61	282	2.61

Continued

Z	Name	r	n	E	n_p
26	Fe	1.4119	2.85	413	2.75
27	Co	1.3846	3.20	424	2.96
28	Ni	1.378	3.34	428	3.05
29	Cu	1.4119	3.03	336	2.84
30	Zn	1.5394	2.00	130	2.42
31	Ga	1.6723	1.33	271	1.98
32	Ge	1.7534	1.05	372	1.72
33	As	1.7316	1.14	285.3	1.87
37	Rb	2.8088	0.00	82.2	2.02
38	Sr	2.3728	0.17	166	1.71
39	Y	1.9924	0.56	422	1.38
40	Zr	1.7746	1.10	603	1.58
41	Nb	1.6237	1.79	730	2.02
42	Mo	1.5504	2.31	658	2.40
43	Tc	1.5043	2.72	661	2.69
44	Ru	1.4873	2.92	650	2.83
45	Rh	1.4873	2.95	554	2.84
46	Pd	1.5224	2.65	376	2.63
47	Ag	1.5982	2.09	284	2.35
48	Cd	1.7316	1.37	112	2.26
49	In	1.8394	1.00	243	1.86
50	Sn	1.8627	0.94	303	1.74
51	Sb	1.9393	0.75	265	1.71
52	Te	2.0105	0.61	215	1.75
55	Cs	3.0275	0.00	77.6	2.05
56	Ba	2.4912	0.14	183	1.64
57	La	2.0748	0.53	431	1.35
58	Ce	2.0168	0.63	417	1.43
59	Pr	2.0207	0.63	357	1.50
60	Nd	2.0138	0.65	328	1.55
62	Sm	1.9924	0.70	206	1.80
63	Eu	2.2551	0.32	179	1.72
64	Gd	1.991	0.71	400	1.50
65	Tb	1.9705	0.76	391	1.54
66	Dy	1.9612	0.79	294	1.68
67	Ho	1.9525	0.82	302	1.68

Continued

Z	Name	r	n	E	n_p
68	Er	1.9425	0.85	317	1.68
69	Tm	1.9301	0.89	233	1.83
70	Yb	2.1427	0.47	154	1.86
71	Lu	1.9177	0.93	428	1.60
72	Hf	1.7534	1.55	621	1.88
73	Ta	1.6288	2.33	782	2.40
74	W	1.5573	2.97	859	2.88
75	Re	1.5196	3.39	775	3.18
76	Os	1.4951	3.71	788	3.42
77	Ir	1.5015	3.66	670	3.35
78	Pt	1.5336	3.29	564	3.07
79	Au	1.5932	2.71	368	2.67
80	Hg	1.8037	1.39	65	2.38
81	Tl	1.8962	1.05	182	2.00
82	Pb	1.9359	0.94	196	1.92
83	Bi	2.0363	0.70	210	1.79
84	Po	2.0799	0.62	144	1.94
90	Th	1.9906	0.83	598	1.40

Table B.14 Fe-X

Z	Name	r	n	E	n_P
26	Fe	1.4119	2.85	413	2.85
27	Co	1.3846	3.20	424	3.03
28	Ni	1.378	3.34	428	3.10
29	Cu	1.4119	3.03	336	2.93
30	Zn	1.5394	2.00	130	2.65
31	Ga	1.6723	1.33	271	2.25
32	Ge	1.7534	1.05	372	1.99
33	As	1.7316	1.14	285.3	2.15
37	Rb	2.8088	0.00	82.2	2.37
38	Sr	2.3728	0.17	166	2.08
39	Y	1.9924	0.56	422	1.69
40	Zr	1.7746	1.10	603	1.81
41	Nb	1.6237	1.79	730	2.17
42	Mo	1.5504	2.31	658	2.51
43	Tc	1.5043	2.72	661	2.77

Z	Name	r	n	E	n_P
44	Ru	1.4873	2.92	650	2.89
45	Rh	1.4873	2.95	554	2.91
46	Pd	1.5224	2.65	376	2.75
47	Ag	1.5982	2.09	284	2.54
48	Cd	1.7316	1.37	112	2.53
49	In	1.8394	1.00	243	2.16
50	Sn	1.8627	0.94	303	2.04
51	Sb	1.9393	0.75	265	2.03
52	Te	2.0105	0.61	215	2.08
55	Cs	3.0275	0.00	77.6	2.40
56	Ba	2.4912	0.14	183	2.02
57	La	2.0748	0.53	431	1.66
58	Ce	2.0168	0.63	417	1.73
59	Pr	2.0207	0.63	357	1.82
60	Nd	2.0138	0.65	328	1.87
62	Sm	1.9924	0.70	206	2.13
63	Eu	2.2551	0.32	179	2.08
64	Gd	1.991	0.71	400	1.80
65	Tb	1.9705	0.76	391	1.83
66	Dy	1.9612	0.79	294	1.99
67	Ho	1.9525	0.82	302	1.99
68	Er	1.9425	0.85	317	1.98
69	Tm	1.9301	0.89	233	2.14
70	Yb	2.1427	0.47	154	2.20
71	Lu	1.9177	0.93	428	1.87
72	Hf	1.7534	1.55	621	2.07
73	Ta	1.6288	2.33	782	2.51
74	W	1.5573	2.97	859	2.93
75	Re	1.5196	3.39	775	3.20
76	Os	1.4951	3.71	788	3.42
77	Ir	1.5015	3.66	670	3.35
78	Pt	1.5336	3.29	564	3.11
79	Au	1.5932	2.71	368	2.78
80	Hg	1.8037	1.39	65	2.65
81	Tl	1.8962	1.05	182	2.30
82	Pb	1.9359	0.94	196	2.23
83	Bi	2.0363	0.70	210	2.12

Z	Name	r	n	E	n_P
84	Po	2.0799	0.62	144	2.27
90	Th	1.9906	0.83	598	1.65

Table B.15 Co-X

Z	Name	r	n	E	n_P
27	Co	1.3846	3.20	424	3.20
28	Ni	1.378	3.34	428	3.27
29	Cu	1.4119	3.03	336	3.12
30	Zn	1.5394	2.00	130	2.92
31	Ga	1.6723	1.33	271	2.47
32	Ge	1.7534	1.05	372	2.19
33	As	1.7316	1.14	285.3	2.37
37	Rb	2.8088	0.00	82.2	2.68
38	Sr	2.3728	0.17	166	2.35
39	Y	1.9924	0.56	422	1.88
40	Zr	1.7746	1.10	603	1.97
41	Nb	1.6237	1.79	730	2.31
42	Mo	1.5504	2.31	658	2.66
43	Tc	1.5043	2.72	661	2.91
44	Ru	1.4873	2.92	650	3.03
45	Rh	1.4873	2.95	554	3.06
46	Pd	1.5224	2.65	376	2.94
47	Ag	1.5982	2.09	284	2.75
48	Cd	1.7316	1.37	112	2.82
49	In	1.8394	1.00	243	2.40
50	Sn	1.8627	0.94	303	2.26
51	Sb	1.9393	0.75	265	2.26
52	Te	2.0105	0.61	215	2.33
55	Cs	3.0275	0.00	77.6	2.70
56	Ba	2.4912	0.14	183	2.28
57	La	2.0748	0.53	431	1.85
58	Ce	2.0168	0.63	417	1.93
59	Pr	2.0207	0.63	357	2.02
60	Nd	2.0138	0.65	328	2.09
62	Sm	1.9924	0.70	206	2.38
63	Eu	2.2551	0.32	179	2.35
64	Gd	1.991	0.71	400	1.99

Continued

Z	Name	r	n	E	n_P
65	Tb	1.9705	0.76	391	2.03
66	Dy	1.9612	0.79	294	2.21
67	Ho	1.9525	0.82	302	2.21
68	Er	1.9425	0.85	317	2.19
69	Tm	1.9301	0.89	233	2.38
70	Yb	2.1427	0.47	154	2.47
71	Lu	1.9177	0.93	428	2.06
72	Hf	1.7534	1.55	621	2.22
73	Ta	1.6288	2.33	782	2.63
74	W	1.5573	2.97	859	3.04
75	Re	1.5196	3.39	775	3.32
76	Os	1.4951	3.71	788	3.53
77	Ir	1.5015	3.66	670	3.48
78	Pt	1.5336	3.29	564	3.25
79	Au	1.5932	2.71	368	2.97
80	Hg	1.8037	1.39	65	2.96
81	Tl	1.8962	1.05	182	2.55
82	Pb	1.9359	0.94	196	2.48
83	Bi	2.0363	0.70	210	2.37
84	Po	2.0799	0.62	144	2.54
90	Th	1.9906	0.83	598	1.81

Table B.16 Ni-X

Z	Name	r	n	E	n_P
28	Ni	1.378	3.34	428	3.34
29	Cu	1.4119	3.03	336	3.20
30	Zn	1.5394	2.00	130	3.03
31	Ga	1.6723	1.33	271	2.56
32	Ge	1.7534	1.05	372	2.28
33	As	1.7316	1.14	285.3	2.46
37	Rb	2.8088	0.00	82.2	2.80
38	Sr	2.3728	0.17	166	2.46
39	Y	1.9924	0.56	422	1.96
40	Zr	1.7746	1.10	603	2.03
41	Nb	1.6237	1.79	730	2.37
42	Mo	1.5504	2.31	658	2.71

Continued

Z	Name	r	n	E	n_P
43	Tc	1.5043	2.72	661	2.97
44	Ru	1.4873	2.92	650	3.09
45	Rh	1.4873	2.95	554	3.12
46	Pd	1.5224	2.65	376	3.02
47	Ag	1.5982	2.09	284	2.84
48	Cd	1.7316	1.37	112	2.93
49	In	1.8394	1.00	243	2.49
50	Sn	1.8627	0.94	303	2.35
51	Sb	1.9393	0.75	265	2.35
52	Te	2.0105	0.61	215	2.43
55	Cs	3.0275	0.00	77.6	2.83
56	Ba	2.4912	0.14	183	2.38
57	La	2.0748	0.53	431	1.93
58	Ce	2.0168	0.63	417	2.00
59	Pr	2.0207	0.63	357	2.11
60	Nd	2.0138	0.65	328	2.17
62	Sm	1.9924	0.70	206	2.48
63	Eu	2.2551	0.32	179	2.45
64	Gd	1.991	0.71	400	2.07
65	Tb	1.9705	0.76	391	2.11
66	Dy	1.9612	0.79	294	2.30
67	Ho	1.9525	0.82	302	2.30
68	Er	1.9425	0.85	317	2.28
69	Tm	1.9301	0.89	233	2.48
70	Yb	2.1427	0.47	154	2.58
71	Lu	1.9177	0.93	428	2.14
72	Hf	1.7534	1.55	621	2.28
73	Ta	1.6288	2.33	782	2.69
74	W	1.5573	2.97	859	3.09
75	Re	1.5196	3.39	775	3.38
76	Os	1.4951	3.71	788	3.58
77	Ir	1.5015	3.66	670	3.53
78	Pt	1.5336	3.29	564	3.32
79	Au	1.5932	2.71	368	3.05
80	Hg	1.8037	1.39	65	3.09
81	Tl	1.8962	1.05	182	2.66

Continued

Z	Name	r	n	E	n_P
82	Pb	1.9359	0.94	196	2.59
83	Bi	2.0363	0.70	210	2.47
84	Po	2.0799	0.62	144	2.66
90	Th	1.9906	0.83	598	1.88

Table B.17 Cu-X

Z	Name	r	n	E	n_P
29	Cu	1.4119	3.03	336	3.03
30	Zn	1.5394	2.00	130	2.74
31	Ga	1.6723	1.33	271	2.27
32	Ge	1.7534	1.05	372	1.99
33	As	1.7316	1.14	285.3	2.16
37	Rb	2.8088	0.00	82.2	2.43
38	Sr	2.3728	0.17	166	2.08
39	Y	1.9924	0.56	422	1.65
40	Zr	1.7746	1.10	603	1.79
41	Nb	1.6237	1.79	730	2.18
42	Mo	1.5504	2.31	658	2.55
43	Tc	1.5043	2.72	661	2.83
44	Ru	1.4873	2.92	650	2.96
45	Rh	1.4873	2.95	554	2.98
46	Pd	1.5224	2.65	376	2.83
47	Ag	1.5982	2.09	284	2.60
48	Cd	1.7316	1.37	112	2.61
49	In	1.8394	1.00	243	2.17
50	Sn	1.8627	0.94	303	2.04
51	Sb	1.9393	0.75	265	2.02
52	Te	2.0105	0.61	215	2.08
55	Cs	3.0275	0.00	77.6	2.46
56	Ba	2.4912	0.14	183	2.01
57	La	2.0748	0.53	431	1.62
58	Ce	2.0168	0.63	417	1.70
59	Pr	2.0207	0.63	357	1.79
60	Nd	2.0138	0.65	328	1.85
62	Sm	1.9924	0.70	206	2.14
63	Eu	2.2551	0.32	179	2.09

Continued

Z	Name	r	n	E	n_P
64	Gd	1.991	0.71	400	1.77
65	Tb	1.9705	0.76	391	1.81
66	Dy	1.9612	0.79	294	1.98
67	Ho	1.9525	0.82	302	1.98
68	Er	1.9425	0.85	317	1.97
69	Tm	1.9301	0.89	233	2.15
70	Yb	2.1427	0.47	154	2.22
71	Lu	1.9177	0.93	428	1.85
72	Hf	1.7534	1.55	621	2.07
73	Ta	1.6288	2.33	782	2.54
74	W	1.5573	2.97	859	2.99
75	Re	1.5196	3.39	775	3.28
76	Os	1.4951	3.71	788	3.51
77	Ir	1.5015	3.66	670	3.45
78	Pt	1.5336	3.29	564	3.19
79	Au	1.5932	2.71	368	2.86
80	Hg	1.8037	1.39	65	2.76
81	Tl	1.8962	1.05	182	2.33
82	Pb	1.9359	0.94	196	2.26
83	Bi	2.0363	0.70	210	2.13
84	Po	2.0799	0.62	144	2.30
90	Th	1.9906	0.83	598	1.62

Table B.18 Zn-X

Z	Name	r	n	E	n_P
30	Zn	1.5394	2.00	130	2.00
31	Ga	1.6723	1.33	271	1.55
32	Ge	1.7534	1.05	372	1.29
33	As	1.7316	1.14	285.3	1.41
37	Rb	2.8088	0.00	82.2	1.23
38	Sr	2.3728	0.17	166	0.98
39	Y	1.9924	0.56	422	0.90
40	Zr	1.7746	1.10	603	1.26
41	Nb	1.6237	1.79	730	1.82
42	Mo	1.5504	2.31	658	2.26
43	Tc	1.5043	2.72	661	2.60

Continued

Z	Name	r	n	E	n_P
44	Ru	1.4873	2.92	650	2.77
45	Rh	1.4873	2.95	554	2.77
46	Pd	1.5224	2.65	376	2.49
47	Ag	1.5982	2.09	284	2.06
48	Cd	1.7316	1.37	112	1.71
49	In	1.8394	1.00	243	1.35
50	Sn	1.8627	0.94	303	1.26
51	Sb	1.9393	0.75	265	1.16
52	Te	2.0105	0.61	215	1.14
55	Cs	3.0275	0.00	77.6	1.25
56	Ba	2.4912	0.14	183	0.92
57	La	2.0748	0.53	431	0.87
58	Ce	2.0168	0.63	417	0.96
59	Pr	2.0207	0.63	357	1.00
60	Nd	2.0138	0.65	328	1.03
62	Sm	1.9924	0.70	206	1.20
63	Eu	2.2551	0.32	179	1.03
64	Gd	1.991	0.71	400	1.03
65	Tb	1.9705	0.76	391	1.07
66	Dy	1.9612	0.79	294	1.16
67	Ho	1.9525	0.82	302	1.17
68	Er	1.9425	0.85	317	1.18
69	Tm	1.9301	0.89	233	1.29
70	Yb	2.1427	0.47	154	1.17
71	Lu	1.9177	0.93	428	1.18
72	Hf	1.7534	1.55	621	1.63
73	Ta	1.6288	2.33	782	2.28
74	W	1.5573	2.97	859	2.84
75	Re	1.5196	3.39	775	3.19
76	Os	1.4951	3.71	788	3.47
77	Ir	1.5015	3.66	670	3.39
78	Pt	1.5336	3.29	564	3.05
79	Au	1.5932	2.71	368	2.53
80	Hg	1.8037	1.39	65	1.80
81	Tl	1.8962	1.05	182	1.45
82	Pb	1.9359	0.94	196	1.36

Continued

Z	Name	r	n	E	n_P
83	Bi	2.0363	0.70	210	1.20
84	Po	2.0799	0.62	144	1.28
90	Th	1.9906	0.83	598	1.04

Table B.19 Ga-X

Z	Name	r	n	E	n_P
31	Ga	1.6723	1.33	271	1.33
32	Ge	1.7534	1.05	372	1.17
33	As	1.7316	1.14	285.3	1.23
37	Rb	2.8088	0.00	82.2	1.02
38	Sr	2.3728	0.17	166	0.89
39	Y	1.9924	0.56	422	0.86
40	Zr	1.7746	1.10	603	1.17
41	Nb	1.6237	1.79	730	1.67
42	Mo	1.5504	2.31	658	2.02
43	Tc	1.5043	2.72	661	2.32
44	Ru	1.4873	2.92	650	2.45
45	Rh	1.4873	2.95	554	2.42
46	Pd	1.5224	2.65	376	2.10
47	Ag	1.5982	2.09	284	1.72
48	Cd	1.7316	1.37	112	1.34
49	In	1.8394	1.00	243	1.17
50	Sn	1.8627	0.94	303	1.12
51	Sb	1.9393	0.75	265	1.04
52	Te	2.0105	0.61	215	1.01
55	Cs	3.0275	0.00	77.6	1.03
56	Ba	2.4912	0.14	183	0.85
57	La	2.0748	0.53	431	0.84
58	Ce	2.0168	0.63	417	0.91
59	Pr	2.0207	0.63	357	0.93
60	Nd	2.0138	0.65	328	0.96
62	Sm	1.9924	0.70	206	1.06
63	Eu	2.2551	0.32	179	0.93
64	Gd	1.991	0.71	400	0.96
65	Tb	1.9705	0.76	391	0.99
66	Dy	1.9612	0.79	294	1.05

Continued

Z	Name	r	n	E	n_P
67	Ho	1.9525	0.82	302	1.06
68	Er	1.9425	0.85	317	1.07
69	Tm	1.9301	0.89	233	1.12
70	Yb	2.1427	0.47	154	1.02
71	Lu	1.9177	0.93	428	1.09
72	Hf	1.7534	1.55	621	1.48
73	Ta	1.6288	2.33	782	2.07
74	W	1.5573	2.97	859	2.58
75	Re	1.5196	3.39	775	2.86
76	Os	1.4951	3.71	788	3.10
77	Ir	1.5015	3.66	670	2.99
78	Pt	1.5336	3.29	564	2.66
79	Au	1.5932	2.71	368	2.13
80	Hg	1.8037	1.39	65	1.34
81	Tl	1.8962	1.05	182	1.22
82	Pb	1.9359	0.94	196	1.16
83	Bi	2.0363	0.70	210	1.05
84	Po	2.0799	0.62	144	1.08
90	Th	1.9906	0.83	598	0.99

Table B.20 Ge-X

Z	Name	r	n	E	n_P
32	Ge	1.7534	1.05	372	1.05
33	As	1.7316	1.14	285.3	1.09
37	Rb	2.8088	0.00	82.2	0.86
38	Sr	2.3728	0.17	166	0.78
39	Y	1.9924	0.56	422	0.79
40	Zr	1.7746	1.10	603	1.08
41	Nb	1.6237	1.79	730	1.54
42	Mo	1.5504	2.31	658	1.85
43	Tc	1.5043	2.72	661	2.12
44	Ru	1.4873	2.92	650	2.24
45	Rh	1.4873	2.95	554	2.19
46	Pd	1.5224	2.65	376	1.85
47	Ag	1.5982	2.09	284	1.50
48	Cd	1.7316	1.37	112	1.12

Continued

Z	Name	r	n	E	n_P
49	In	1.8394	1.00	243	1.03
50	Sn	1.8627	0.94	303	1.00
51	Sb	1.9393	0.75	265	0.92
52	Te	2.0105	0.61	215	0.89
55	Cs	3.0275	0.00	77.6	0.87
56	Ba	2.4912	0.14	183	0.75
57	La	2.0748	0.53	431	0.77
58	Ce	2.0168	0.63	417	0.83
59	Pr	2.0207	0.63	357	0.84
60	Nd	2.0138	0.65	328	0.86
62	Sm	1.9924	0.70	206	0.92
63	Eu	2.2551	0.32	179	0.81
64	Gd	1.991	0.71	400	0.87
65	Tb	1.9705	0.76	391	0.90
66	Dy	1.9612	0.79	294	0.93
67	Ho	1.9525	0.82	302	0.94
68	Er	1.9425	0.85	317	0.96
69	Tm	1.9301	0.89	233	0.99
70	Yb	2.1427	0.47	154	0.88
71	Lu	1.9177	0.93	428	0.99
72	Hf	1.7534	1.55	621	1.36
73	Ta	1.6288	2.33	782	1.92
74	W	1.5573	2.97	859	2.39
75	Re	1.5196	3.39	775	2.63
76	Os	1.4951	3.71	788	2.86
77	Ir	1.5015	3.66	670	2.72
78	Pt	1.5336	3.29	564	2.40
79	Au	1.5932	2.71	368	1.88
80	Hg	1.8037	1.39	65	1.10
81	Tl	1.8962	1.05	182	1.05
82	Pb	1.9359	0.94	196	1.01
83	Bi	2.0363	0.70	210	0.92
84	Po	2.0799	0.62	144	0.93
90	Th	1.9906	0.83	598	0.91

Table B.21 As-X

Z	Name	r	n	E	n_P
33	As	1.7316	1.14	285.3	1.14
37	Rb	2.8088	0.00	82.2	0.88
38	Sr	2.3728	0.17	166	0.78
39	Y	1.9924	0.56	422	0.79
40	Zr	1.7746	1.10	603	1.11
41	Nb	1.6237	1.79	730	1.61
42	Mo	1.5504	2.31	658	1.95
43	Tc	1.5043	2.72	661	2.25
44	Ru	1.4873	2.92	650	2.38
45	Rh	1.4873	2.95	554	2.34
46	Pd	1.5224	2.65	376	2.00
47	Ag	1.5982	2.09	284	1.61
48	Cd	1.7316	1.37	112	1.21
49	In	1.8394	1.00	243	1.07
50	Sn	1.8627	0.94	303	1.03
51	Sb	1.9393	0.75	265	0.95
52	Te	2.0105	0.61	215	0.91
55	Cs	3.0275	0.00	77.6	0.90
56	Ba	2.4912	0.14	183	0.75
57	La	2.0748	0.53	431	0.77
58	Ce	2.0168	0.63	417	0.84
59	Pr	2.0207	0.63	357	0.86
60	Nd	2.0138	0.65	328	0.88
62	Sm	1.9924	0.70	206	0.95
63	Eu	2.2551	0.32	179	0.82
64	Gd	1.991	0.71	400	0.89
65	Tb	1.9705	0.76	391	0.92
66	Dy	1.9612	0.79	294	0.96
67	Ho	1.9525	0.82	302	0.97
68	Er	1.9425	0.85	317	0.99
69	Tm	1.9301	0.89	233	1.02
70	Yb	2.1427	0.47	154	0.91
71	Lu	1.9177	0.93	428	1.01
72	Hf	1.7534	1.55	621	1.42
73	Ta	1.6288	2.33	782	2.01
74	W	1.5573	2.97	859	2.51

Continued

Z	Name	r	n	E	n_P
75	Re	1.5196	3.39	775	2.79
76	Os	1.4951	3.71	788	3.03
77	Ir	1.5015	3.66	670	2.90
78	Pt	1.5336	3.29	564	2.57
79	Au	1.5932	2.71	368	2.03
80	Hg	1.8037	1.39	65	1.19
81	Tl	1.8962	1.05	182	1.11
82	Pb	1.9359	0.94	196	1.06
83	Bi	2.0363	0.70	210	0.95
84	Po	2.0799	0.62	144	0.96
90	Th	1.9906	0.83	598	0.93

Table B.22 Rb-X

Z	Name	r	n	E	n_P
37	Rb	2.8088	0.07	82.2	0.07
38	Sr	2.3728	0.17	166	0.14
39	Y	1.9924	0.56	422	0.48
40	Zr	1.7746	1.10	603	0.98
41	Nb	1.6237	1.79	730	1.62
42	Mo	1.5504	2.31	658	2.06
43	Tc	1.5043	2.72	661	2.43
44	Ru	1.4873	2.92	650	2.60
45	Rh	1.4873	2.95	554	2.58
46	Pd	1.5224	2.65	376	2.19
47	Ag	1.5982	2.09	284	1.63
48	Cd	1.7316	1.37	112	0.82
49	In	1.8394	1.00	243	0.76
50	Sn	1.8627	0.94	303	0.75
51	Sb	1.9393	0.75	265	0.59
52	Te	2.0105	0.61	215	0.46
55	Cs	3.0275	0.07	77.6	0.07
56	Ba	2.4912	0.14	183	0.12
57	La	2.0748	0.53	431	0.45
58	Ce	2.0168	0.63	417	0.54
59	Pr	2.0207	0.63	357	0.52
60	Nd	2.0138	0.65	328	0.53

Continued

Z	Name	r	n	E	n_p
62	Sm	1.9924	0.70	206	0.52
63	Eu	2.2551	0.32	179	0.24
64	Gd	1.991	0.71	400	0.60
65	Tb	1.9705	0.76	391	0.64
66	Dy	1.9612	0.79	294	0.63
67	Ho	1.9525	0.82	302	0.66
68	Er	1.9425	0.85	317	0.69
69	Tm	1.9301	0.89	233	0.67
70	Yb	2.1427	0.47	154	0.33
71	Lu	1.9177	0.93	428	0.79
72	Hf	1.7534	1.55	621	1.38
73	Ta	1.6288	2.33	782	2.11
74	W	1.5573	2.97	859	2.72
75	Re	1.5196	3.39	775	3.07
76	Os	1.4951	3.71	788	3.37
77	Ir	1.5015	3.66	670	3.26
78	Pt	1.5336	3.29	564	2.88
79	Au	1.5932	2.71	368	2.23
80	Hg	1.8037	1.39	65	0.65
81	Tl	1.8962	1.05	182	0.75
82	Pb	1.9359	0.94	196	0.68
83	Bi	2.0363	0.70	210	0.52
84	Po	2.0799	0.62	144	0.42
90	Th	1.9906	0.83	598	0.74

Table B.23 Sr-X

Z	Name	r	n	E	n_p
38	Sr	2.3728	0.17	166	0.17
39	Y	1.9924	0.56	422	0.45
40	Zr	1.7746	1.10	603	0.90
41	Nb	1.6237	1.79	730	1.49
42	Mo	1.5504	2.31	658	1.88
43	Tc	1.5043	2.72	661	2.21
44	Ru	1.4873	2.92	650	2.36
45	Rh	1.4873	2.95	554	2.31
46	Pd	1.5224	2.65	376	1.89

Continued

Z	Name	r	n	E	n_P
47	Ag	1.5982	2.09	284	1.38
48	Cd	1.7316	1.37	112	0.66
49	In	1.8394	1.00	243	0.66
50	Sn	1.8627	0.94	303	0.66
51	Sb	1.9393	0.75	265	0.53
52	Te	2.0105	0.61	215	0.42
55	Cs	3.0275	0.07	77.6	0.14
56	Ba	2.4912	0.14	183	0.16
57	La	2.0748	0.53	431	0.43
58	Ce	2.0168	0.63	417	0.50
59	Pr	2.0207	0.63	357	0.48
60	Nd	2.0138	0.65	328	0.49
62	Sm	1.9924	0.70	206	0.46
63	Eu	2.2551	0.32	179	0.25
64	Gd	1.991	0.71	400	0.55
65	Tb	1.9705	0.76	391	0.59
66	Dy	1.9612	0.79	294	0.57
67	Ho	1.9525	0.82	302	0.59
68	Er	1.9425	0.85	317	0.62
69	Tm	1.9301	0.89	233	0.59
70	Yb	2.1427	0.47	154	0.32
71	Lu	1.9177	0.93	428	0.72
72	Hf	1.7534	1.55	621	1.26
73	Ta	1.6288	2.33	782	1.95
74	W	1.5573	2.97	859	2.52
75	Re	1.5196	3.39	775	2.82
76	Os	1.4951	3.71	788	3.10
77	Ir	1.5015	3.66	670	2.96
78	Pt	1.5336	3.29	564	2.58
79	Au	1.5932	2.71	368	1.92
80	Hg	1.8037	1.39	65	0.51
81	Tl	1.8962	1.05	182	0.63
82	Pb	1.9359	0.94	196	0.59
83	Bi	2.0363	0.70	210	0.47
84	Po	2.0799	0.62	144	0.38
90	Th	1.9906	0.83	598	0.69

Table B.24 Y-X

Z	Name	r	n	E	n_p
39	Y	1.9924	0.56	422	0.56
40	Zr	1.7746	1.10	603	0.88
41	Nb	1.6237	1.79	730	1.34
42	Mo	1.5504	2.31	658	1.62
43	Tc	1.5043	2.72	661	1.88
44	Ru	1.4873	2.92	650	1.99
45	Rh	1.4873	2.95	554	1.92
46	Pd	1.5224	2.65	376	1.55
47	Ag	1.5982	2.09	284	1.17
48	Cd	1.7316	1.37	112	0.73
49	In	1.8394	1.00	243	0.72
51	Sb	1.9393	0.75	265	0.63
52	Te	2.0105	0.61	215	0.58
55	Cs	3.0275	0.07	77.6	0.48
56	Ba	2.4912	0.14	183	0.43
57	La	2.0748	0.53	431	0.54
58	Ce	2.0168	0.63	417	0.60
59	Pr	2.0207	0.63	357	0.59
60	Nd	2.0138	0.65	328	0.60
62	Sm	1.9924	0.70	206	0.61
63	Eu	2.2551	0.32	179	0.49
64	Gd	1.991	0.71	400	0.64
65	Tb	1.9705	0.76	391	0.66
66	Dy	1.9612	0.79	294	0.66
67	Ho	1.9525	0.82	302	0.67
68	Er	1.9425	0.85	317	0.68
69	Tm	1.9301	0.89	233	0.68
70	Yb	2.1427	0.47	154	0.54
71	Lu	1.9177	0.93	428	0.75
72	Hf	1.7534	1.55	621	1.15
73	Ta	1.6288	2.33	782	1.71
74	W	1.5573	2.97	859	2.18
75	Re	1.5196	3.39	775	2.39
76	Os	1.4951	3.71	788	2.61
77	Ir	1.5015	3.66	670	2.46
78	Pt	1.5336	3.29	564	2.12

Continued

Z	Name	r	n	E	n_p
79	Au	1.5932	2.71	368	1.56
80	Hg	1.8037	1.39	65	0.67
81	Tl	1.8962	1.05	182	0.71
82	Pb	1.9359	0.94	196	0.68
83	Bi	2.0363	0.70	210	0.61
84	Po	2.0799	0.62	144	0.58
90	Th	1.9906	0.83	598	0.72

Table B.25 Zr-X

Z	Name	r	n	E	n_P
40	Zr	1.7746	1.10	603	1.10
41	Nb	1.6237	1.79	730	1.48
42	Mo	1.5504	2.31	658	1.73
43	Tc	1.5043	2.72	661	1.95
44	Ru	1.4873	2.92	650	2.04
45	Rh	1.4873	2.95	554	1.99
46	Pd	1.5224	2.65	376	1.70
47	Ag	1.5982	2.09	284	1.41
48	Cd	1.7316	1.37	112	1.14
49	In	1.8394	1.00	243	1.07
50	Sn	1.8627	0.94	303	1.04
51	Sb	1.9393	0.75	265	0.99
52	Te	2.0105	0.61	215	0.97
55	Cs	3.0275	0.07	77.6	0.98
56	Ba	2.4912	0.14	183	0.88
57	La	2.0748	0.53	431	0.86
58	Ce	2.0168	0.63	417	0.91
59	Pr	2.0207	0.63	357	0.92
60	Nd	2.0138	0.65	328	0.94
62	Sm	1.9924	0.70	206	1.00
63	Eu	2.2551	0.32	179	0.92
64	Gd	1.991	0.71	400	0.95
65	Tb	1.9705	0.76	391	0.97
66	Dy	1.9612	0.79	294	1.00
67	Ho	1.9525	0.82	302	1.01
68	Er	1.9425	0.85	317	1.01

Continued

Z	Name	r	n	E	n_P
69	Tm	1.9301	0.89	233	1.04
70	Yb	2.1427	0.47	154	0.97
71	Lu	1.9177	0.93	428	1.03
72	Hf	1.7534	1.55	621	1.33
73	Ta	1.6288	2.33	782	1.79
74	W	1.5573	2.97	859	2.20
75	Re	1.5196	3.39	775	2.39
76	Os	1.4951	3.71	788	2.58
77	Ir	1.5015	3.66	670	2.44
78	Pt	1.5336	3.29	564	2.16
79	Au	1.5932	2.71	368	1.71
80	Hg	1.8037	1.39	65	1.13
81	Tl	1.8962	1.05	182	1.09
82	Pb	1.9359	0.94	196	1.06
83	Bi	2.0363	0.70	210	1.00
84	Po	2.0799	0.62	144	1.01
90	Th	1.9906	0.83	598	0.97

Table B.26 Nb-X

Z	Name	r	n	E	n_P
41	Nb	1.6237	1.79	730	1.79
42	Mo	1.5504	2.31	658	2.04
43	Tc	1.5043	2.72	661	2.23
44	Ru	1.4873	2.92	650	2.32
45	Rh	1.4873	2.95	554	2.29
46	Pd	1.5224	2.65	376	2.08
47	Ag	1.5982	2.09	284	1.87
48	Cd	1.7316	1.37	112	1.74
49	In	1.8394	1.00	243	1.59
50	Sn	1.8627	0.94	303	1.54
51	Sb	1.9393	0.75	265	1.51
52	Te	2.0105	0.61	215	1.52
55	Cs	3.0275	0.07	77.6	1.63
56	Ba	2.4912	0.14	183	1.46
57	La	2.0748	0.53	431	1.32
58	Ce	2.0168	0.63	417	1.37

Continued

Z	Name	r	n	E	n_P
59	Pr	2.0207	0.63	357	1.41
60	Nd	2.0138	0.65	328	1.44
62	Sm	1.9924	0.70	206	1.55
63	Eu	2.2551	0.32	179	1.50
64	Gd	1.991	0.71	400	1.41
65	Tb	1.9705	0.76	391	1.43
66	Dy	1.9612	0.79	294	1.50
67	Ho	1.9525	0.82	302	1.51
68	Er	1.9425	0.85	317	1.51
69	Tm	1.9301	0.89	233	1.57
70	Yb	2.1427	0.47	154	1.56
71	Lu	1.9177	0.93	428	1.47
72	Hf	1.7534	1.55	621	1.68
73	Ta	1.6288	2.33	782	2.07
74	W	1.5573	2.97	859	2.43
75	Re	1.5196	3.39	775	2.62
76	Os	1.4951	3.71	788	2.79
77	Ir	1.5015	3.66	670	2.68
78	Pt	1.5336	3.29	564	2.45
79	Au	1.5932	2.71	368	2.10
80	Hg	1.8037	1.39	65	1.76
81	Tl	1.8962	1.05	182	1.64
82	Pb	1.9359	0.94	196	1.61
83	Bi	2.0363	0.70	210	1.55
84	Po	2.0799	0.62	144	1.60
90	Th	1.9906	0.83	598	1.36

Table B.27 Mo-X

Z	Name	r	n	E	n_P
42	Mo	1.5504	2.31	658	2.31
43	Tc	1.5043	2.72	661	2.52
44	Ru	1.4873	2.92	650	2.61
45	Rh	1.4873	2.95	554	2.60
46	Pd	1.5224	2.65	376	2.43
47	Ag	1.5982	2.09	284	2.24
48	Cd	1.7316	1.37	112	2.17

Continued

Z	Name	r	n	E	n_P
49	In	1.8394	1.00	243	1.95
50	Sn	1.8627	0.94	303	1.87
51	Sb	1.9393	0.75	265	1.86
52	Te	2.0105	0.61	215	1.89
55	Cs	3.0275	0.07	77.6	2.07
56	Ba	2.4912	0.14	183	1.84
57	La	2.0748	0.53	431	1.60
58	Ce	2.0168	0.63	417	1.66
59	Pr	2.0207	0.63	357	1.72
60	Nd	2.0138	0.65	328	1.75
62	Sm	1.9924	0.70	206	1.92
63	Eu	2.2551	0.32	179	1.88
64	Gd	1.991	0.71	400	1.70
65	Tb	1.9705	0.76	391	1.73
66	Dy	1.9612	0.79	294	1.84
67	Ho	1.9525	0.82	302	1.84
68	Er	1.9425	0.85	317	1.83
69	Tm	1.9301	0.89	233	1.93
70	Yb	2.1427	0.47	154	1.96
71	Lu	1.9177	0.93	428	1.76
72	Hf	1.7534	1.55	621	1.94
73	Ta	1.6288	2.33	782	2.32
74	W	1.5573	2.97	859	2.68
75	Re	1.5196	3.39	775	2.89
76	Os	1.4951	3.71	788	3.07
77	Ir	1.5015	3.66	670	2.99
78	Pt	1.5336	3.29	564	2.76
79	Au	1.5932	2.71	368	2.45
80	Hg	1.8037	1.39	65	2.22
81	Tl	1.8962	1.05	182	2.03
82	Pb	1.9359	0.94	196	1.99
83	Bi	2.0363	0.70	210	1.92
84	Po	2.0799	0.62	144	2.00
90	Th	1.9906	0.83	598	1.60

Table B.28　Tc-X

Z	Name	r	n	E	n_P
43	Tc	1.5043	2.72	661	2.72
44	Ru	1.4873	2.92	650	2.82
45	Rh	1.4873	2.95	554	2.83
46	Pd	1.5224	2.65	376	2.70
47	Ag	1.5982	2.09	284	2.53
48	Cd	1.7316	1.37	112	2.53
49	In	1.8394	1.00	243	2.26
50	Sn	1.8627	0.94	303	2.16
51	Sb	1.9393	0.75	265	2.16
52	Te	2.0105	0.61	215	2.20
55	Cs	3.0275	0.07	77.6	2.44
56	Ba	2.4912	0.14	183	2.16
57	La	2.0748	0.53	431	1.86
58	Ce	2.0168	0.63	417	1.91
59	Pr	2.0207	0.63	357	1.99
60	Nd	2.0138	0.65	328	2.03
62	Sm	1.9924	0.70	206	2.24
63	Eu	2.2551	0.32	179	2.21
64	Gd	1.991	0.71	400	1.97
65	Tb	1.9705	0.76	391	1.99
66	Dy	1.9612	0.79	294	2.13
67	Ho	1.9525	0.82	302	2.13
68	Er	1.9425	0.85	317	2.12
69	Tm	1.9301	0.89	233	2.24
70	Yb	2.1427	0.47	154	2.30
71	Lu	1.9177	0.93	428	2.02
72	Hf	1.7534	1.55	621	2.16
73	Ta	1.6288	2.33	782	2.51
74	W	1.5573	2.97	859	2.86
75	Re	1.5196	3.39	775	3.08
76	Os	1.4951	3.71	788	3.26
77	Ir	1.5015	3.66	670	3.19
78	Pt	1.5336	3.29	564	2.99
79	Au	1.5932	2.71	368	2.72
80	Hg	1.8037	1.39	65	2.60
81	Tl	1.8962	1.05	182	2.36

Continued

Z	Name	r	n	E	n_P
82	Pb	1.9359	0.94	196	2.32
83	Bi	2.0363	0.70	210	2.24
84	Po	2.0799	0.62	144	2.35
90	Th	1.9906	0.83	598	1.82

Table B.29 Ru-X

Z	Name	r	n	E	n_P
44	Ru	1.4873	2.92	650	2.92
45	Rh	1.4873	2.95	554	2.94
46	Pd	1.5224	2.65	376	2.82
47	Ag	1.5982	2.09	284	2.67
48	Cd	1.7316	1.37	112	2.69
49	In	1.8394	1.00	243	2.40
50	Sn	1.8627	0.94	303	2.29
51	Sb	1.9393	0.75	265	2.29
52	Te	2.0105	0.61	215	2.35
55	Cs	3.0275	0.07	77.6	2.62
56	Ba	2.4912	0.14	183	2.31
57	La	2.0748	0.53	431	1.97
58	Ce	2.0168	0.63	417	2.03
59	Pr	2.0207	0.63	357	2.11
60	Nd	2.0138	0.65	328	2.16
62	Sm	1.9924	0.70	206	2.39
63	Eu	2.2551	0.32	179	2.36
64	Gd	1.991	0.71	400	2.08
65	Tb	1.9705	0.76	391	2.11
66	Dy	1.9612	0.79	294	2.26
67	Ho	1.9525	0.82	302	2.25
68	Er	1.9425	0.85	317	2.24
69	Tm	1.9301	0.89	233	2.38
70	Yb	2.1427	0.47	154	2.45
71	Lu	1.9177	0.93	428	2.13
72	Hf	1.7534	1.55	621	2.25
73	Ta	1.6288	2.33	782	2.60
74	W	1.5573	2.97	859	2.95
75	Re	1.5196	3.39	775	3.18

Continued

Z	Name	r	n	E	n_P
76	Os	1.4951	3.71	788	3.35
77	Ir	1.5015	3.66	670	3.29
78	Pt	1.5336	3.29	564	3.09
79	Au	1.5932	2.71	368	2.85
80	Hg	1.8037	1.39	65	2.78
81	Tl	1.8962	1.05	182	2.51
82	Pb	1.9359	0.94	196	2.46
83	Bi	2.0363	0.70	210	2.38
84	Po	2.0799	0.62	144	2.50
90	Th	1.9906	0.83	598	1.92

Table B.30 Rh-X

Z	Name	r	n	E	n_P
45	Rh	1.4873	2.95	554	2.95
46	Pd	1.5224	2.65	376	2.83
47	Ag	1.5982	2.09	284	2.66
48	Cd	1.7316	1.37	112	2.69
49	In	1.8394	1.00	243	2.36
50	Sn	1.8627	0.94	303	2.24
51	Sb	1.9393	0.75	265	2.24
52	Te	2.0105	0.61	215	2.30
55	Cs	3.0275	0.07	77.6	2.60
56	Ba	2.4912	0.14	183	2.26
57	La	2.0748	0.53	431	1.89
58	Ce	2.0168	0.63	417	1.96
59	Pr	2.0207	0.63	357	2.04
60	Nd	2.0138	0.65	328	2.10
62	Sm	1.9924	0.70	206	2.34
63	Eu	2.2551	0.32	179	2.31
64	Gd	1.991	0.71	400	2.01
65	Tb	1.9705	0.76	391	2.05
66	Dy	1.9612	0.79	294	2.20
67	Ho	1.9525	0.82	302	2.20
68	Er	1.9425	0.85	317	2.19
69	Tm	1.9301	0.89	233	2.34
70	Yb	2.1427	0.47	154	2.41

Continued

Z	Name	r	n	E	n_P
71	Lu	1.9177	0.93	428	2.07
72	Hf	1.7534	1.55	621	2.21
73	Ta	1.6288	2.33	782	2.59
74	W	1.5573	2.97	859	2.96
75	Re	1.5196	3.39	775	3.21
76	Os	1.4951	3.71	788	3.40
77	Ir	1.5015	3.66	670	3.34
78	Pt	1.5336	3.29	564	3.13
79	Au	1.5932	2.71	368	2.86
80	Hg	1.8037	1.39	65	2.79
81	Tl	1.8962	1.05	182	2.48
82	Pb	1.9359	0.94	196	2.43
83	Bi	2.0363	0.70	210	2.33
84	Po	2.0799	0.62	144	2.47
90	Th	1.9906	0.83	598	1.85

Table B.31 Pd-X

Z	Name	r	n	E	n_P
46	Pd	1.5224	2.65	376	2.65
47	Ag	1.5982	2.09	284	2.41
48	Cd	1.7316	1.37	112	2.36
49	In	1.8394	1.00	243	2.00
50	Sn	1.8627	0.94	303	1.89
51	Sb	1.9393	0.75	265	1.87
52	Te	2.0105	0.61	215	1.91
55	Cs	3.0275	0.07	77.6	2.21
56	Ba	2.4912	0.14	183	1.83
57	La	2.0748	0.53	431	1.52
58	Ce	2.0168	0.63	417	1.59
59	Pr	2.0207	0.63	357	1.67
60	Nd	2.0138	0.65	328	1.72
62	Sm	1.9924	0.70	206	1.96
63	Eu	2.2551	0.32	179	1.90
64	Gd	1.991	0.71	400	1.65
65	Tb	1.9705	0.76	391	1.69
66	Dy	1.9612	0.79	294	1.84

Continued

Z	Name	r	n	E	n_P
67	Ho	1.9525	0.82	302	1.83
68	Er	1.9425	0.85	317	1.83
69	Tm	1.9301	0.89	233	1.98
70	Yb	2.1427	0.47	154	2.02
71	Lu	1.9177	0.93	428	1.74
72	Hf	1.7534	1.55	621	1.97
73	Ta	1.6288	2.33	782	2.43
74	W	1.5573	2.97	859	2.87
75	Re	1.5196	3.39	775	3.15
76	Os	1.4951	3.71	788	3.37
77	Ir	1.5015	3.66	670	3.29
78	Pt	1.5336	3.29	564	3.04
79	Au	1.5932	2.71	368	2.68
80	Hg	1.8037	1.39	65	2.47
81	Tl	1.8962	1.05	182	2.13
82	Pb	1.9359	0.94	196	2.07
83	Bi	2.0363	0.70	210	1.95
84	Po	2.0799	0.62	144	2.09
90	Th	1.9906	0.83	598	1.53

Table B.32 Ag-X

Z	Name	r	n	E	n_P
47	Ag	1.5982	2.09	284	2.09
48	Cd	1.7316	1.37	112	1.88
49	In	1.8394	1.00	243	1.58
50	Sn	1.8627	0.94	303	1.49
51	Sb	1.9393	0.75	265	1.44
52	Te	2.0105	0.61	215	1.45
55	Cs	3.0275	0.07	77.6	1.65
56	Ba	2.4912	0.14	183	1.32
57	La	2.0748	0.53	431	1.15
58	Ce	2.0168	0.63	417	1.22
59	Pr	2.0207	0.63	357	1.27
60	Nd	2.0138	0.65	328	1.31
62	Sm	1.9924	0.70	206	1.50
63	Eu	2.2551	0.32	179	1.40

Continued

Z	Name	r	n	E	n_P
64	Gd	1.991	0.71	400	1.28
65	Tb	1.9705	0.76	391	1.32
66	Dy	1.9612	0.79	294	1.43
67	Ho	1.9525	0.82	302	1.43
68	Er	1.9425	0.85	317	1.43
69	Tm	1.9301	0.89	233	1.54
70	Yb	2.1427	0.47	154	1.52
71	Lu	1.9177	0.93	428	1.39
72	Hf	1.7534	1.55	621	1.72
73	Ta	1.6288	2.33	782	2.26
74	W	1.5573	2.97	859	2.75
75	Re	1.5196	3.39	775	3.04
76	Os	1.4951	3.71	788	3.28
77	Ir	1.5015	3.66	670	3.19
78	Pt	1.5336	3.29	564	2.89
79	Au	1.5932	2.71	368	2.44
80	Hg	1.8037	1.39	65	1.96
81	Tl	1.8962	1.05	182	1.68
82	Pb	1.9359	0.94	196	1.62
83	Bi	2.0363	0.70	210	1.50
84	Po	2.0799	0.62	144	1.59
90	Th	1.9906	0.83	598	1.23

Table B.33 Cd-X

Z	Name	r	n	E	n_P
48	Cd	1.7316	1.37	112	1.37
49	In	1.8394	1.00	243	1.11
50	Sn	1.8627	0.94	303	1.05
51	Sb	1.9393	0.75	265	0.94
52	Te	2.0105	0.61	215	0.87
55	Cs	3.0275	0.07	77.6	0.84
56	Ba	2.4912	0.14	183	0.61
57	La	2.0748	0.53	431	0.70
58	Ce	2.0168	0.63	417	0.79
59	Pr	2.0207	0.63	357	0.81
60	Nd	2.0138	0.65	328	0.83

Continued

Z	Name	r	n	E	n_P
62	Sm	1.9924	0.70	206	0.94
63	Eu	2.2551	0.32	179	0.73
64	Gd	1.991	0.71	400	0.86
65	Tb	1.9705	0.76	391	0.90
66	Dy	1.9612	0.79	294	0.95
67	Ho	1.9525	0.82	302	0.97
68	Er	1.9425	0.85	317	0.99
69	Tm	1.9301	0.89	233	1.04
70	Yb	2.1427	0.47	154	0.85
71	Lu	1.9177	0.93	428	1.02
72	Hf	1.7534	1.55	621	1.53
73	Ta	1.6288	2.33	782	2.21
74	W	1.5573	2.97	859	2.78
75	Re	1.5196	3.39	775	3.14
76	Os	1.4951	3.71	788	3.42
77	Ir	1.5015	3.66	670	3.33
78	Pt	1.5336	3.29	564	2.98
79	Au	1.5932	2.71	368	2.40
80	Hg	1.8037	1.39	65	1.38
81	Tl	1.8962	1.05	182	1.18
82	Pb	1.9359	0.94	196	1.10
83	Bi	2.0363	0.70	210	0.94
84	Po	2.0799	0.62	144	0.95
90	Th	1.9906	0.83	598	0.92

Table B.34 In-X

Z	Name	r	n	E	n_P
49	In	1.8394	1.00	243	1.00
50	Sn	1.8627	0.94	303	0.96
51	Sb	1.9393	0.75	265	0.87
52	Te	2.0105	0.61	215	0.81
55	Cs	3.0275	0.07	77.6	0.77
56	Ba	2.4912	0.14	183	0.63
57	La	2.0748	0.53	431	0.70
58	Ce	2.0168	0.63	417	0.77
59	Pr	2.0207	0.63	357	0.78

Continued

Z	Name	r	n	E	n_P
60	Nd	2.0138	0.65	328	0.80
62	Sm	1.9924	0.70	206	0.86
63	Eu	2.2551	0.32	179	0.71
64	Gd	1.991	0.71	400	0.82
65	Tb	1.9705	0.76	391	0.85
66	Dy	1.9612	0.79	294	0.88
67	Ho	1.9525	0.82	302	0.90
68	Er	1.9425	0.85	317	0.91
69	Tm	1.9301	0.89	233	0.94
70	Yb	2.1427	0.47	154	0.79
71	Lu	1.9177	0.93	428	0.95
72	Hf	1.7534	1.55	621	1.40
73	Ta	1.6288	2.33	782	2.01
74	W	1.5573	2.97	859	2.53
75	Re	1.5196	3.39	775	2.82
76	Os	1.4951	3.71	788	3.07
77	Ir	1.5015	3.66	670	2.95
78	Pt	1.5336	3.29	564	2.60
79	Au	1.5932	2.71	368	2.03
80	Hg	1.8037	1.39	65	1.08
81	Tl	1.8962	1.05	182	1.02
82	Pb	1.9359	0.94	196	0.97
83	Bi	2.0363	0.70	210	0.86
84	Po	2.0799	0.62	144	0.86
90	Th	1.9906	0.83	598	0.88

Table B.35 Sn-X

Z	Name	r	n	E	n_P
50	Sn	1.8627	0.94	303	0.94
51	Sb	1.9393	0.75	265	0.85
52	Te	2.0105	0.61	215	0.80
55	Cs	3.0275	0.07	77.6	0.76
56	Ba	2.4912	0.14	183	0.64
57	La	2.0748	0.53	431	0.70
58	Ce	2.0168	0.63	417	0.76
59	Pr	2.0207	0.63	357	0.77

Continued

Z	Name	r	n	E	n_P
60	Nd	2.0138	0.65	328	0.79
62	Sm	1.9924	0.70	206	0.84
63	Eu	2.2551	0.32	179	0.71
64	Gd	1.991	0.71	400	0.81
65	Tb	1.9705	0.76	391	0.84
66	Dy	1.9612	0.79	294	0.86
67	Ho	1.9525	0.82	302	0.88
68	Er	1.9425	0.85	317	0.89
69	Tm	1.9301	0.89	233	0.91
70	Yb	2.1427	0.47	154	0.78
71	Lu	1.9177	0.93	428	0.93
72	Hf	1.7534	1.55	621	1.35
73	Ta	1.6288	2.33	782	1.94
74	W	1.5573	2.97	859	2.44
75	Re	1.5196	3.39	775	2.70
76	Os	1.4951	3.71	788	2.94
77	Ir	1.5015	3.66	670	2.81
78	Pt	1.5336	3.29	564	2.47
79	Au	1.5932	2.71	368	1.91
80	Hg	1.8037	1.39	65	1.02
81	Tl	1.8962	1.05	182	0.98
82	Pb	1.9359	0.94	196	0.94
83	Bi	2.0363	0.70	210	0.84
84	Po	2.0799	0.62	144	0.83
90	Th	1.9906	0.83	598	0.87

Table B.36 Sb-X

Z	Name	r	n	E	n_P
51	Sb	1.9393	0.75	265	0.75
52	Te	2.0105	0.61	215	0.69
55	Cs	3.0275	0.07	77.6	0.60
56	Ba	2.4912	0.14	183	0.50
57	La	2.0748	0.53	431	0.61
58	Ce	2.0168	0.63	417	0.68
59	Pr	2.0207	0.63	357	0.68
60	Nd	2.0138	0.65	328	0.69

Z	Name	r	n	E	n_P
62	Sm	1.9924	0.70	206	0.73
63	Eu	2.2551	0.32	179	0.58
64	Gd	1.991	0.71	400	0.73
65	Tb	1.9705	0.76	391	0.76
66	Dy	1.9612	0.79	294	0.77
67	Ho	1.9525	0.82	302	0.79
68	Er	1.9425	0.85	317	0.80
69	Tm	1.9301	0.89	233	0.81
70	Yb	2.1427	0.47	154	0.65
71	Lu	1.9177	0.93	428	0.86
72	Hf	1.7534	1.55	621	1.31
73	Ta	1.6288	2.33	782	1.93
74	W	1.5573	2.97	859	2.45
75	Re	1.5196	3.39	775	2.72
76	Os	1.4951	3.71	788	2.97
77	Ir	1.5015	3.66	670	2.83
78	Pt	1.5336	3.29	564	2.48
79	Au	1.5932	2.71	368	1.89
80	Hg	1.8037	1.39	65	0.88
81	Tl	1.8962	1.05	182	0.87
82	Pb	1.9359	0.94	196	0.83
83	Bi	2.0363	0.70	210	0.73
84	Po	2.0799	0.62	144	0.70
90	Th	1.9906	0.83	598	0.81

Table B.37 Te-X

Z	Name	r	n	E	n_P
52	Te	2.0105	0.61	215	0.61
55	Cs	3.0275	0.07	77.6	0.47
56	Ba	2.4912	0.14	183	0.40
57	La	2.0748	0.53	431	0.56
58	Ce	2.0168	0.63	417	0.62
59	Pr	2.0207	0.63	357	0.62
60	Nd	2.0138	0.65	328	0.63
62	Sm	1.9924	0.70	206	0.66
63	Eu	2.2551	0.32	179	0.48
64	Gd	1.991	0.71	400	0.68

Continued

Z	Name	r	n	E	n_P
65	Tb	1.9705	0.76	391	0.71
66	Dy	1.9612	0.79	294	0.72
67	Ho	1.9525	0.82	302	0.73
68	Er	1.9425	0.85	317	0.75
69	Tm	1.9301	0.89	233	0.75
70	Yb	2.1427	0.47	154	0.55
71	Lu	1.9177	0.93	428	0.82
72	Hf	1.7534	1.55	621	1.31
73	Ta	1.6288	2.33	782	1.96
74	W	1.5573	2.97	859	2.50
75	Re	1.5196	3.39	775	2.79
76	Os	1.4951	3.71	788	3.05
77	Ir	1.5015	3.66	670	2.92
78	Pt	1.5336	3.29	564	2.55
79	Au	1.5932	2.71	368	1.94
80	Hg	1.8037	1.39	65	0.79
81	Tl	1.8962	1.05	182	0.81
82	Pb	1.9359	0.94	196	0.77
83	Bi	2.0363	0.70	210	0.66
84	Po	2.0799	0.62	144	0.61
90	Th	1.9906	0.83	598	0.77

Table B.38 Cs-X

Z	Name	r	n	E	n_P
55	Cs	3.0275	0.07	77.6	0.07
56	Ba	2.4912	0.14	183	0.12
57	La	2.0748	0.53	431	0.46
58	Ce	2.0168	0.63	417	0.54
59	Pr	2.0207	0.63	357	0.53
60	Nd	2.0138	0.65	328	0.54
62	Sm	1.9924	0.70	206	0.53
63	Eu	2.2551	0.32	179	0.25
64	Gd	1.991	0.71	400	0.61
65	Tb	1.9705	0.76	391	0.65
66	Dy	1.9612	0.79	294	0.64
67	Ho	1.9525	0.82	302	0.66

Continued

Z	Name	r	n	E	n_P
68	Er	1.9425	0.85	317	0.70
69	Tm	1.9301	0.89	233	0.68
70	Yb	2.1427	0.47	154	0.34
71	Lu	1.9177	0.93	428	0.80
72	Hf	1.7534	1.55	621	1.39
73	Ta	1.6288	2.33	782	2.12
74	W	1.5573	2.97	859	2.73
75	Re	1.5196	3.39	775	3.09
76	Os	1.4951	3.71	788	3.39
77	Ir	1.5015	3.66	670	3.28
78	Pt	1.5336	3.29	564	2.90
79	Au	1.5932	2.71	368	2.25
80	Hg	1.8037	1.39	65	0.67
81	Tl	1.8962	1.05	182	0.76
82	Pb	1.9359	0.94	196	0.69
83	Bi	2.0363	0.70	210	0.53
84	Po	2.0799	0.62	144	0.43
90	Th	1.9906	0.83	598	0.74

Table B.39 Ba-X

Z	Name	r	n	E	n_P
56	Ba	2.4912	0.14	183	0.14
57	La	2.0748	0.53	431	0.41
58	Ce	2.0168	0.63	417	0.48
59	Pr	2.0207	0.63	357	0.46
60	Nd	2.0138	0.65	328	0.47
62	Sm	1.9924	0.70	206	0.44
63	Eu	2.2551	0.32	179	0.23
64	Gd	1.991	0.71	400	0.53
65	Tb	1.9705	0.76	391	0.57
66	Dy	1.9612	0.79	294	0.54
67	Ho	1.9525	0.82	302	0.56
68	Er	1.9425	0.85	317	0.59
69	Tm	1.9301	0.89	233	0.56
70	Yb	2.1427	0.47	154	0.29
71	Lu	1.9177	0.93	428	0.70

Z	Name	r	n	E	n_P
72	Hf	1.7534	1.55	621	1.23
73	Ta	1.6288	2.33	782	1.91
74	W	1.5573	2.97	859	2.47
75	Re	1.5196	3.39	775	2.77
76	Os	1.4951	3.71	788	3.04
77	Ir	1.5015	3.66	670	2.90
78	Pt	1.5336	3.29	564	2.52
79	Au	1.5932	2.71	368	1.86
80	Hg	1.8037	1.39	65	0.47
81	Tl	1.8962	1.05	182	0.60
82	Pb	1.9359	0.94	196	0.56
83	Bi	2.0363	0.70	210	0.44
84	Po	2.0799	0.62	144	0.35
90	Th	1.9906	0.83	598	0.67

Table B.40 La-X

Z	Name	r	n	E	n_P
57	La	2.0748	0.53	431	0.53
58	Ce	2.0168	0.63	417	0.58
59	Pr	2.0207	0.63	357	0.57
60	Nd	2.0138	0.65	328	0.58
62	Sm	1.9924	0.70	206	0.58
63	Eu	2.2551	0.32	179	0.47
64	Gd	1.991	0.71	400	0.62
65	Tb	1.9705	0.76	391	0.64
66	Dy	1.9612	0.79	294	0.63
67	Ho	1.9525	0.82	302	0.65
68	Er	1.9425	0.85	317	0.66
69	Tm	1.9301	0.89	233	0.65
70	Yb	2.1427	0.47	154	0.51
71	Lu	1.9177	0.93	428	0.73
72	Hf	1.7534	1.55	621	1.13
73	Ta	1.6288	2.33	782	1.69
74	W	1.5573	2.97	859	2.15
75	Re	1.5196	3.39	775	2.37
76	Os	1.4951	3.71	788	2.59
77	Ir	1.5015	3.66	670	2.43

Continued

Z	Name	r	n	E	n_P
78	Pt	1.5336	3.29	564	2.10
79	Au	1.5932	2.71	368	1.53
80	Hg	1.8037	1.39	65	0.64
81	Tl	1.8962	1.05	182	0.68
82	Pb	1.9359	0.94	196	0.66
83	Bi	2.0363	0.70	210	0.58
84	Po	2.0799	0.62	144	0.55
90	Th	1.9906	0.83	598	0.70

Table B.41 Ce-X

Z	Name	r	n	E	n_P
58	Ce	2.0168	0.63	417	0.63
59	Pr	2.0207	0.63	357	0.63
60	Nd	2.0138	0.65	328	0.64
62	Sm	1.9924	0.70	206	0.65
63	Eu	2.2551	0.32	179	0.54
64	Gd	1.991	0.71	400	0.67
65	Tb	1.9705	0.76	391	0.70
66	Dy	1.9612	0.79	294	0.70
67	Ho	1.9525	0.82	302	0.71
68	Er	1.9425	0.85	317	0.73
69	Tm	1.9301	0.89	233	0.72
70	Yb	2.1427	0.47	154	0.59
71	Lu	1.9177	0.93	428	0.78
72	Hf	1.7534	1.55	621	1.18
73	Ta	1.6288	2.33	782	1.74
74	W	1.5573	2.97	859	2.21
75	Re	1.5196	3.39	775	2.43
76	Os	1.4951	3.71	788	2.65
77	Ir	1.5015	3.66	670	2.50
78	Pt	1.5336	3.29	564	2.16
79	Au	1.5932	2.71	368	1.61
80	Hg	1.8037	1.39	65	0.73
81	Tl	1.8962	1.05	182	0.76
82	Pb	1.9359	0.94	196	0.73
83	Bi	2.0363	0.70	210	0.65

Continued

Z	Name	r	n	E	n_P
84	Po	2.0799	0.62	144	0.63
90	Th	1.9906	0.83	598	0.75

Table B.42 Pr-X

Z	Name	r	n	E	n_P
59	Pr	2.0207	0.63	357	0.63
60	Nd	2.0138	0.65	328	0.64
62	Sm	1.9924	0.70	206	0.66
63	Eu	2.2551	0.32	179	0.53
64	Gd	1.991	0.71	400	0.67
65	Tb	1.9705	0.76	391	0.70
66	Dy	1.9612	0.79	294	0.70
67	Ho	1.9525	0.82	302	0.72
68	Er	1.9425	0.85	317	0.73
69	Tm	1.9301	0.89	233	0.73
70	Yb	2.1427	0.47	154	0.58
71	Lu	1.9177	0.93	428	0.79
72	Hf	1.7534	1.55	621	1.22
73	Ta	1.6288	2.33	782	1.80
74	W	1.5573	2.97	859	2.28
75	Re	1.5196	3.39	775	2.52
76	Os	1.4951	3.71	788	2.75
77	Ir	1.5015	3.66	670	2.60
78	Pt	1.5336	3.29	564	2.26
79	Au	1.5932	2.71	368	1.69
80	Hg	1.8037	1.39	65	0.75
81	Tl	1.8962	1.05	182	0.77
82	Pb	1.9359	0.94	196	0.74
83	Bi	2.0363	0.70	210	0.66
84	Po	2.0799	0.62	144	0.63
90	Th	1.9906	0.83	598	0.76

Table B.43 Nd-X

Z	Name	r	n	E	n_P
60	Nd	2.0138	0.65	328	0.65
62	Sm	1.9924	0.70	206	0.67

Continued

Z	Name	r	n	E	n_P
63	Eu	2.2551	0.32	179	0.53
64	Gd	1.991	0.71	400	0.68
65	Tb	1.9705	0.76	391	0.71
66	Dy	1.9612	0.79	294	0.72
67	Ho	1.9525	0.82	302	0.73
68	Er	1.9425	0.85	317	0.75
69	Tm	1.9301	0.89	233	0.75
70	Yb	2.1427	0.47	154	0.59
71	Lu	1.9177	0.93	428	0.81
72	Hf	1.7534	1.55	621	1.24
73	Ta	1.6288	2.33	782	1.83
74	W	1.5573	2.97	859	2.33
75	Re	1.5196	3.39	775	2.58
76	Os	1.4951	3.71	788	2.81
77	Ir	1.5015	3.66	670	2.67
78	Pt	1.5336	3.29	564	2.32
79	Au	1.5932	2.71	368	1.74
80	Hg	1.8037	1.39	65	0.77
81	Tl	1.8962	1.05	182	0.79
82	Pb	1.9359	0.94	196	0.76
83	Bi	2.0363	0.70	210	0.67
84	Po	2.0799	0.62	144	0.64
90	Th	1.9906	0.83	598	0.77

Table B.44 Sm-X

Z	Name	r	n	E	n_P
62	Sm	1.9924	0.70	206	0.70
63	Eu	2.2551	0.32	179	0.53
64	Gd	1.991	0.71	400	0.71
65	Tb	1.9705	0.76	391	0.74
66	Dy	1.9612	0.79	294	0.75
67	Ho	1.9525	0.82	302	0.77
68	Er	1.9425	0.85	317	0.79
69	Tm	1.9301	0.89	233	0.80
70	Yb	2.1427	0.47	154	0.60
71	Lu	1.9177	0.93	428	0.86

Continued

Z	Name	r	n	E	n_P
72	Hf	1.7534	1.55	621	1.34
73	Ta	1.6288	2.33	782	1.99
74	W	1.5573	2.97	859	2.53
75	Re	1.5196	3.39	775	2.83
76	Os	1.4951	3.71	788	3.09
77	Ir	1.5015	3.66	670	2.96
78	Pt	1.5336	3.29	564	2.60
79	Au	1.5932	2.71	368	1.99
80	Hg	1.8037	1.39	65	0.87
81	Tl	1.8962	1.05	182	0.87
82	Pb	1.9359	0.94	196	0.82
83	Bi	2.0363	0.70	210	0.70
84	Po	2.0799	0.62	144	0.67
90	Th	1.9906	0.83	598	0.80

Table B.45 Eu-X

Z	Name	r	n	E	n_P
63	Eu	2.2551	0.32	179	0.32
64	Gd	1.991	0.71	400	0.59
65	Tb	1.9705	0.76	391	0.63
66	Dy	1.9612	0.79	294	0.61
67	Ho	1.9525	0.82	302	0.63
68	Er	1.9425	0.85	317	0.66
69	Tm	1.9301	0.89	233	0.64
70	Yb	2.1427	0.47	154	0.39
71	Lu	1.9177	0.93	428	0.75
72	Hf	1.7534	1.55	621	1.28
73	Ta	1.6288	2.33	782	1.95
74	W	1.5573	2.97	859	2.51
75	Re	1.5196	3.39	775	2.82
76	Os	1.4951	3.71	788	3.09
77	Ir	1.5015	3.66	670	2.95
78	Pt	1.5336	3.29	564	2.58
79	Au	1.5932	2.71	368	1.93
80	Hg	1.8037	1.39	65	0.61
81	Tl	1.8962	1.05	182	0.69

Continued

Z	Name	r	n	E	n_P
82	Pb	1.9359	0.94	196	0.65
83	Bi	2.0363	0.70	210	0.53
84	Po	2.0799	0.62	144	0.46
90	Th	1.9906	0.83	598	0.71

Table B.46 Gd-X

Z	Name	r	n	E	n_P
64	Gd	1.991	0.71	400	0.71
65	Tb	1.9705	0.76	391	0.74
66	Dy	1.9612	0.79	294	0.75
67	Ho	1.9525	0.82	302	0.76
68	Er	1.9425	0.85	317	0.77
69	Tm	1.9301	0.89	233	0.78
70	Yb	2.1427	0.47	154	0.65
71	Lu	1.9177	0.93	428	0.83
72	Hf	1.7534	1.55	621	1.22
73	Ta	1.6288	2.33	782	1.78
74	W	1.5573	2.97	859	2.25
75	Re	1.5196	3.39	775	2.48
76	Os	1.4951	3.71	788	2.70
77	Ir	1.5015	3.66	670	2.56
78	Pt	1.5336	3.29	564	2.22
79	Au	1.5932	2.71	368	1.67
80	Hg	1.8037	1.39	65	0.81
81	Tl	1.8962	1.05	182	0.82
82	Pb	1.9359	0.94	196	0.79
83	Bi	2.0363	0.70	210	0.71
84	Po	2.0799	0.62	144	0.69
90	Th	1.9906	0.83	598	0.78

Table B.47 Tb-X

Z	Name	r	n	E	n_P
65	Tb	1.9705	0.76	391	0.76
66	Dy	1.9612	0.79	294	0.78
67	Ho	1.9525	0.82	302	0.79
68	Er	1.9425	0.85	317	0.80

Continued

Z	Name	r	n	E	n_P
69	Tm	1.9301	0.89	233	0.81
70	Yb	2.1427	0.47	154	0.68
71	Lu	1.9177	0.93	428	0.85
72	Hf	1.7534	1.55	621	1.25
73	Ta	1.6288	2.33	782	1.81
74	W	1.5573	2.97	859	2.28
75	Re	1.5196	3.39	775	2.51
76	Os	1.4951	3.71	788	2.74
77	Ir	1.5015	3.66	670	2.59
78	Pt	1.5336	3.29	564	2.26
79	Au	1.5932	2.71	368	1.71
80	Hg	1.8037	1.39	65	0.85
81	Tl	1.8962	1.05	182	0.86
82	Pb	1.9359	0.94	196	0.82
83	Bi	2.0363	0.70	210	0.74
84	Po	2.0799	0.62	144	0.73
90	Th	1.9906	0.83	598	0.80

Table B.48 Dy-X

Z	Name	r	n	E	n_P
66	Dy	1.9612	0.79	294	0.79
67	Ho	1.9525	0.82	302	0.80
68	Er	1.9425	0.85	317	0.82
69	Tm	1.9301	0.89	233	0.83
70	Yb	2.1427	0.47	154	0.68
71	Lu	1.9177	0.93	428	0.87
72	Hf	1.7534	1.55	621	1.31
73	Ta	1.6288	2.33	782	1.91
74	W	1.5573	2.97	859	2.41
75	Re	1.5196	3.39	775	2.68
76	Os	1.4951	3.71	788	2.92
77	Ir	1.5015	3.66	670	2.78
78	Pt	1.5336	3.29	564	2.44
79	Au	1.5932	2.71	368	1.86
80	Hg	1.8037	1.39	65	0.90

Z	Name	r	n	E	n_P
81	Tl	1.8962	1.05	182	0.89
82	Pb	1.9359	0.94	196	0.85
83	Bi	2.0363	0.70	210	0.75
84	Po	2.0799	0.62	144	0.73
90	Th	1.9906	0.83	598	0.82

Table B.49 Ho-X

Z	Name	r	n	E	n_P
67	Ho	1.9525	0.82	302	0.82
68	Er	1.9425	0.85	317	0.83
69	Tm	1.9301	0.89	233	0.85
70	Yb	2.1427	0.47	154	0.70
71	Lu	1.9177	0.93	428	0.88
72	Hf	1.7534	1.55	621	1.31
73	Ta	1.6288	2.33	782	1.91
74	W	1.5573	2.97	859	2.41
75	Re	1.5196	3.39	775	2.67
76	Os	1.4951	3.71	788	2.91
77	Ir	1.5015	3.66	670	2.77
78	Pt	1.5336	3.29	564	2.43
79	Au	1.5932	2.71	368	1.86
80	Hg	1.8037	1.39	65	0.92
81	Tl	1.8962	1.05	182	0.91
82	Pb	1.9359	0.94	196	0.87
83	Bi	2.0363	0.70	210	0.77
84	Po	2.0799	0.62	144	0.75
90	Th	1.9906	0.83	598	0.83

Table B.50 Er-X

Z	Name	r	n	E	n_P
68	Er	1.9425	0.85	317	0.85
69	Tm	1.9301	0.89	233	0.86
70	Yb	2.1427	0.47	154	0.73
71	Lu	1.9177	0.93	428	0.90
72	Hf	1.7534	1.55	621	1.31
73	Ta	1.6288	2.33	782	1.90
74	W	1.5573	2.97	859	2.40

Continued

Z	Name	r	n	E	n_P
75	Re	1.5196	3.39	775	2.65
76	Os	1.4951	3.71	788	2.89
77	Ir	1.5015	3.66	670	2.75
78	Pt	1.5336	3.29	564	2.41
79	Au	1.5932	2.71	368	1.85
80	Hg	1.8037	1.39	65	0.94
81	Tl	1.8962	1.05	182	0.92
82	Pb	1.9359	0.94	196	0.88
83	Bi	2.0363	0.70	210	0.79
84	Po	2.0799	0.62	144	0.78
90	Th	1.9906	0.83	598	0.84

Table B.51 Tm-X

Z	Name	r	n	E	n_P
69	Tm	1.9301	0.89	233	0.89
70	Yb	2.1427	0.47	154	0.72
71	Lu	1.9177	0.93	428	0.92
72	Hf	1.7534	1.55	621	1.37
73	Ta	1.6288	2.33	782	2.00
74	W	1.5573	2.97	859	2.52
75	Re	1.5196	3.39	775	2.81
76	Os	1.4951	3.71	788	3.07
77	Ir	1.5015	3.66	670	2.94
78	Pt	1.5336	3.29	564	2.59
79	Au	1.5932	2.71	368	2.01
80	Hg	1.8037	1.39	65	1.00
81	Tl	1.8962	1.05	182	0.96
82	Pb	1.9359	0.94	196	0.91
83	Bi	2.0363	0.70	210	0.80
84	Po	2.0799	0.62	144	0.78
90	Th	1.9906	0.83	598	0.85

Table B.52 Yb-X

Z	Name	r	n	E	n_P
70	Yb	2.1427	0.47	154	0.47
71	Lu	1.9177	0.93	428	0.81

Continued

Z	Name	r	n	E	n_P
72	Hf	1.7534	1.55	621	1.34
73	Ta	1.6288	2.33	782	2.02
74	W	1.5573	2.97	859	2.59
75	Re	1.5196	3.39	775	2.91
76	Os	1.4951	3.71	788	3.18
77	Ir	1.5015	3.66	670	3.06
78	Pt	1.5336	3.29	564	2.69
79	Au	1.5932	2.71	368	2.05
80	Hg	1.8037	1.39	65	0.75
81	Tl	1.8962	1.05	182	0.79
82	Pb	1.9359	0.94	196	0.74
83	Bi	2.0363	0.70	210	0.60
84	Po	2.0799	0.62	144	0.54
90	Th	1.9906	0.83	598	0.76

Table B.53 Lu-X

Z	Name	r	n	E	n_P
71	Lu	1.9177	0.93	428	0.93
72	Hf	1.7534	1.55	621	1.30
73	Ta	1.6288	2.33	782	1.83
74	W	1.5573	2.97	859	2.29
75	Re	1.5196	3.39	775	2.52
76	Os	1.4951	3.71	788	2.73
77	Ir	1.5015	3.66	670	2.59
78	Pt	1.5336	3.29	564	2.28
79	Au	1.5932	2.71	368	1.76
80	Hg	1.8037	1.39	65	0.99
81	Tl	1.8962	1.05	182	0.97
82	Pb	1.9359	0.94	196	0.93
83	Bi	2.0363	0.70	210	0.86
84	Po	2.0799	0.62	144	0.85
90	Th	1.9906	0.83	598	0.87

Table B.54 Hf-X

Z	Name	r	n	E	n_P
72	Hf	1.7534	1.55	621	1.55

Continued

Z	Name	r	n	E	n_P
73	Ta	1.6288	2.33	782	1.99
74	W	1.5573	2.97	859	2.37
75	Re	1.5196	3.39	775	2.57
76	Os	1.4951	3.71	788	2.76
77	Ir	1.5015	3.66	670	2.64
78	Pt	1.5336	3.29	564	2.38
79	Au	1.5932	2.71	368	1.98
80	Hg	1.8037	1.39	65	1.54
81	Tl	1.8962	1.05	182	1.44
82	Pb	1.9359	0.94	196	1.41
83	Bi	2.0363	0.70	210	1.34
84	Po	2.0799	0.62	144	1.38
90	Th	1.9906	0.83	598	1.20

Table B.55 Ta-X

Z	Name	r	n	E	n_P
73	Ta	1.6288	2.33	782	2.33
74	W	1.5573	2.97	859	2.66
75	Re	1.5196	3.39	775	2.86
76	Os	1.4951	3.71	788	3.02
77	Ir	1.5015	3.66	670	2.94
78	Pt	1.5336	3.29	564	2.73
79	Au	1.5932	2.71	368	2.45
80	Hg	1.8037	1.39	65	2.26
81	Tl	1.8962	1.05	182	2.09
82	Pb	1.9359	0.94	196	2.05
83	Bi	2.0363	0.70	210	1.98
84	Po	2.0799	0.62	144	2.06
90	Th	1.9906	0.83	598	1.68

Table B.56 W-X

Z	Name	r	n	E	n_P
74	W	1.5573	2.97	859	2.97
75	Re	1.5196	3.39	775	3.17
76	Os	1.4951	3.71	788	3.33
77	Ir	1.5015	3.66	670	3.27

Continued

Z	Name	r	n	E	n_P
78	Pt	1.5336	3.29	564	3.10
79	Au	1.5932	2.71	368	2.89
80	Hg	1.8037	1.39	65	2.86
81	Tl	1.8962	1.05	182	2.63
82	Pb	1.9359	0.94	196	2.59
83	Bi	2.0363	0.70	210	2.52
84	Po	2.0799	0.62	144	2.63
90	Th	1.9906	0.83	598	2.09

Table B.57 Re-X

Z	Name	r	n	E	n_P
75	Re	1.5196	3.39	775	3.39
76	Os	1.4951	3.71	788	3.56
77	Ir	1.5015	3.66	670	3.51
78	Pt	1.5336	3.29	564	3.35
79	Au	1.5932	2.71	368	3.17
80	Hg	1.8037	1.39	65	3.24
81	Tl	1.8962	1.05	182	2.95
82	Pb	1.9359	0.94	196	2.90
83	Bi	2.0363	0.70	210	2.82
84	Po	2.0799	0.62	144	2.96
90	Th	1.9906	0.83	598	2.28

Table B.58 Os-X

Z	Name	r	n	E	n_P
76	Os	1.4951	3.71	788	3.71
77	Ir	1.5015	3.66	670	3.69
78	Pt	1.5336	3.29	564	3.54
79	Au	1.5932	2.71	368	3.40
80	Hg	1.8037	1.39	65	3.54
81	Tl	1.8962	1.05	182	3.21
82	Pb	1.9359	0.94	196	3.16
83	Bi	2.0363	0.70	210	3.08
84	Po	2.0799	0.62	144	3.24
90	Th	1.9906	0.83	598	2.47

Table B.59 Ir-X

Z	Name	r	n	E	n_P
77	Ir	1.5015	3.66	670	3.66
78	Pt	1.5336	3.29	564	3.49
79	Au	1.5932	2.71	368	3.32
80	Hg	1.8037	1.39	65	3.46
81	Tl	1.8962	1.05	182	3.10
82	Pb	1.9359	0.94	196	3.04
83	Bi	2.0363	0.70	210	2.95
84	Po	2.0799	0.62	144	3.12
90	Th	1.9906	0.83	598	2.32

Table B.60 Pt-X

Z	Name	r	n	E	n_P
78	Pt	1.5336	3.29	564	3.29
79	Au	1.5932	2.71	368	3.07
80	Hg	1.8037	1.39	65	3.10
81	Tl	1.8962	1.05	182	2.75
82	Pb	1.9359	0.94	196	2.69
83	Bi	2.0363	0.70	210	2.59
84	Po	2.0799	0.62	144	2.75
90	Th	1.9906	0.83	598	2.03

Table B.61 Au-X

Z	Name	r	n	E	n_P
79	Au	1.5932	2.71	368	2.71
80	Hg	1.8037	1.39	65	2.52
81	Tl	1.8962	1.05	182	2.16
82	Pb	1.9359	0.94	196	2.10
83	Bi	2.0363	0.70	210	1.98
84	Po	2.0799	0.62	144	2.12
90	Th	1.9906	0.83	598	1.55

Table B.62 Tl-X

Z	Name	r	n	E	n_P
81	Tl	1.8962	1.05	182	1.05
82	Pb	1.9359	0.94	196	0.99
83	Bi	2.0363	0.70	210	0.86

Continued

Z	Name	r	n	E	n_P
84	Po	2.0799	0.62	144	0.86
90	Th	1.9906	0.83	598	0.88

Table B.63 Pb-X

Z	Name	r	n	E	n_P
82	Pb	1.9359	0.94	196	0.94
83	Bi	2.0363	0.70	210	0.82
84	Po	2.0799	0.62	144	0.80
90	Th	1.9906	0.83	598	0.86

Table B.64 Bi-X

Z	Name	r	n	E	n_P
83	Bi	2.0363	0.70	210	0.70
84	Po	2.0799	0.62	144	0.67
90	Th	1.9906	0.83	598	0.80

Table B.65 Po-X

Z	Name	r	n	E	n_P
84	Po	2.0799	0.62	144	0.62
90	Th	1.9906	0.83	598	0.79

References

1. Yu Ruihuang. Empirical electron theory of solid and molecules. Chinese Science Bulletin, 1978,23(4): 217
2. Zhang Relin. Empirical electron theory of solid and molecules. Changchun: Jilin Science and Technology Press, 1992
3. Li Shichun. Atomic phase diagram. Progress in Natural Science, 2004,14(2):113
4. Brandes E A, Brook G B. Smithells metals reference book. Bodmin, Cornwall: Butterworth-Heinemann Ltd., 1992:11
5. Li Shichun. Research on superplasticity of Zn-Al eutectic alloy. Beijing: Tsinghua University, 2000
6. Li Shichun. Phase boundary sliding model controlled by diffusion solution zone in superplastic deformation. Chinese Science Bulletin, 2004,49(14):1355;2002,47(14):1228-1232
7. Garay J E, Anselmi-Tamburini U, Munir Z A. Enhanced growth of intermetallic phases in the Ni-Ti system by current effects. Acta. Materialia, 2003,51(15):4487-4495
8. Li Shichun. Relationship between the valence electron theory and the electron density theory in crystal. Progress in Natural Science, 1999, 9(7): 505-511
9. Thomas L H. The calculation of atomic fields. Proc. Camb. Phil. Soc., 1926,23:542
10. Fermi E. Eine statistische methode zur bestimmung einiger eigenschaften des atoms und ihre anwendung auf die theories des periodischen systems der elemente. Z. Phys.,1928,48:73
11. Dirac P A M. Note on exchange phenomena in the thomas atom. Proc. Camb. Phil. Soc.,1930,26: 376
12. Cheng Kaijia, Cheng Suyu. Theoretical foundations of condensed materials. Progress in Natural Science, 1996, 6(1): 12
13. Meyer A, Umar I H, Yound W H. Lattice spacings and compressibilities vs pauling radii and valencies. Phys. Rev. 1971, B4:3287-3291
14. Miedema A R, de Boer F R, de Chatel P F. Empirical description of the role of electronegativity in alloy formation. J. Phys. F: Metal Phys., 1973, 3: 1558
15. Cheng Dayong, Wang Shaoqing, Ye Hengqiang. Calculations showing a correlation between electronic density and bulk modulus in fcc and bcc metals. Phys. Review, 2001, 64B: 24107
16. Cheng K J, Cheng S Y. Application of TFD model and Yu's theory to materials design. Progress in Natural Science, 1993,3(5): 417
17. Vegard L. Die Konstitution der mischkristalle und die raumfllung der atome. Z. Phys., 1921, (5): 17-26
18. Vegard L, Dale H. Untersuchungen ü ber mischkristalle und legierungen. Z. Kristallogr, 1922, 67:

148-162
19. Pines B J. On solid solutions. Journal of Physics, 1940,3(4-5): 309-319
20. Friedel J. Deviations from Vegard's Law. Phys. Mag., 1955, 46: 514-516
21. Zen E-an. Validity of "Vegard's Law". Am. Mineralogist, 1956,41: 523-524
22. Dienes G J. Lattice parameter and short range order. Acta. Met., 1958,41: 278-282
23. Gschneidner K A, Vineyard G H. Departure from Vegard's Law. Journal of Applied Physics, 1962, 33(12): 3444-3450
24. Landa A I, Psakhe S G, Panin V E, et al. About the validity of the Vegard's rule for the solid solutions. Physica, 1981(6): 118-120
25. Hafner J. A note on Vegard's and Zen's laws. J. Phys. F: Met. Phys. 1985: 15, 43-48
26. Denton A R, Ashcroft N W. Vegard's Law. Phys. Rev. 1991,A43: 3161
27. Li Wei, Pessa Markus. Lattice paramter in GaNAs epilayers on GaAs: Deviation from Vegard's Law. Applied Physics Letters, 2001,78(19): 2864
28. Lubarda V A. On the effective lattice parameter of binary alloys. Mechanics of Materials,2003(35): 53-68
29. Underwood E E. A review of superplasticity and related phenomenon. J. Metals. 1962(14):914-919
30. Nadai A, Manjoine M J. High-speed tension tests at elevated temperatures. J. Appl. Mech., 1941,8 (6):A77-A91
31. Backofen W A, Turner I R, Avery D H. Superplasticity in an Al-Zn alloy. Transactions of the ASMA, 1964,57: 980-990
32. Song Yuquan, Lian Shujun. Mechanical theory of strain rate sensitivity index m and new method of measurement using constant load. Materials Science and Technology,1989,5(5):443-449
33. Aravas N, Kim K S, Leckie F A. On the calculations of the stored energy of cold work. J. Eng. Mater. and Tech., 1990,112(10):465-470
34. Bever M B, Holt D L, Titchener A L. The stored energy of cold work. Oxford: Pergamon Press, 1989:23-50
35. Shi G C. Thermomechanics of nonequilibrium and irreversible processes. Advances in Mechanics, 1989,19(2):158-168
36. Ha Kuanfu. Microscopic mechanics of metals. Beijing: Science Press, 1983:198-201
37. Chaudhury P K,Mohamed F A. Effect of impurity content on superplastic flow in the Zn-22%Al alloy. Acta. Metall,1988,36(4):1099-1110
38. Bochvara A A, Sviderskaya Z A. Superplasticity in Zinc-Aluminum alloys. Izv. Akad. Nauk. SSSR, Ordel Tekhn. Nauk,1945(9):821-827
39. Nabarro F R N. Steady state diffusional creep. Phil. Mag., 1976,A16:231-237
40. Backofen W A, Murty G S, Zehr S W. Evidence for diffusional creep with low strain rate sensitivity. Trans. Metall. Soc. AIME. 1968,242(2):329-331
41. Mukherjeea A K. The rate controlling mechanism in superplasticity. Mater. Sci. Eng.,1971(8):83-89
42. Lee D. Structural changes during the superplastic deformation. Met. Trans., 1970,1(1):309-311
43. Giffkins R C.Grain rearrangements during superplastic deformation. J. Mater. Sci.,1978, 13:1926-1936
44. Ball A, Hutchson M M. Superplasticity in the Aluminum-Zinc eutectoid. Met. Sci. J., 1969(3):1-6
45. Ashby M F, Verall R A. Diffusion—accommodated flow and superplasticity.Acta. Metall., 1973,21 (2):149-163

46. Feng Duan. Metal physics, Vol. 1. Beijing: Science Press, 1998: 521-529
47. Sun Lianchao, Tian Rongzhang. Physical metallurgy of zinc and its alloy. Changsha: Press of Central South University of Techology,1994:303-311
48. Hans Loffler. Structure and structure development of Al-Zn alloys. Akademie Verlag,1995:27
49. Wang Chongyu. Energetics of metallic defect and electronic structure of doped grain boundary. Acta. Metallurgica Sinica, 1997,33(1):54-68
50. Cheng Kaijia, Cheng Suyu. Application of the TFD model and Yu's theory to material design. Progress in Natural Science, 1993, 3(3):211-230
51. Brandes E A, Brook G B. Smithells metals reference book. Bodmin, Cornwall: Butterworth-Heinemann Ltd., 1992:11-57
52. Powder diffraction file, International centre for diffraction data, 1901, Parklane, Swarthmore, Pennsylvania 19081-2389,USA. 1989
53. Cahn R W, Haasen P. Physical metallurgy. Amsterdam: North-Holland Physics Publishing,1996: 194
54. Feng Duan. Metal physics, Vol. 1. Beijing: Science Press, 1998: 331-334
55. Xiao Shenxiu, Wang Chongyu, Chen Tianlang. Discrete variational method in density functional theory and its application in chemistry and materials physics. Beijing: Science Press, 1998: 162
56. Ip S W, Toguri J M. The equivalency of surface tension, surface energy and surface free energy. Journal of Materials Science,1994,29:688-692
57. Conrad H, Sprecher A F. The electroplastic effect in metals. In: Nabarro F R N. Dislocations in Solids. Elsevier Science,B.V. 1989,8:497
58. Conrad H, Sprecher A F, Cao W D, et al. Electroplasticity: The effect of electricity on the mechanical properties of metals. JOM,1990(9): 28-33
59. Cao Weidi, Conrad H. Effects of stacking fault energy and temperature on the electroplastic effect in fcc metals. In: Chu S N G, Arsenault R J, Sadananda K, et al. Micromechanics of advanced Materials. Minerals, Metals & Materials Society,1995: 225-236
60. Conrad H, Cao W D, Lu X P, et al. Effect of an electric field on the superplasticity of 7475Al. Scripta Metall, 1989,23:697-702
61. Mabuchi M, Higashi K, Okada Y,et al. Very high strain-rate superplasticity in a particulate Si_3N_4/6061 aluminum composite. Scripta Metall, 1991, 25(11):2517-2522
62. Wu M Y, Wadsworth J, Sherby O D. Elimination of the threshold stress for creep by thermal cycling in Oxide-dispersion-strengthened materials. Scripta Metall,1987,21:1159-1164
63. Hong S H, Sherby O D, Divecha A P,et al. Internal stress superplasticity in 2024 Al-SiC whisker reinforced composites. J. of Composite Materials, 1988,22(2):103-123
64. Li Shichun, Mechanism of rare earth increasing the superplasticity in Al-Zn-Mg Alloys. In: Materials Research Society of China. Proceeding of the 2th National Conference on Materials.Wuhan, 1988: 160-163
65. Zhang Xinyu. Texture of metals and alloys. Beijing: Science Press, 1976
66. Li Shichun,Hu Xiulian,Liu Runchang. Effect of rare earth on superplasticity in Zn-5Al alloy. Heat Treatment of Metals,1995(1):20
67. Zhou Tairui,Hu Shifei. The effect of RE on structure and property of Zn-Al superplastic material. Proc. of the 3rd inter conf. on RE devel. and Applic. Bejing:Metallurgical Industry Press, 1995:90-93

68. Kittel C. Introduction to solid state physics. New York: John Wiley & Sons Inc., 2005

69. Pearson W B. A handbook of lattice spacings and structures of metals and alloys. New York: Pergamon, 1958: 255-889

Acknowledgements and Postscript

Superplasticity was the research area I chose for my master's degree 20 years ago; my thesis was written on the effect of rare earth elements on the superplasticity of Al-Zn-Mg alloys. Later on, while working in UPC, I designed and made a superplastic tensile machine with the help of two students. An elongation of 5000% was obtained for the Zn-5Al eutectic alloy. Henceforth I have been focusing on elucidation of the microscopic mechanism of this 5000% elongation.

Zn-5Al eutectic alloy has a layered microstructures. There is an interfacial zone between the Al and Zn phases, and the thickness of this interfacial zone can be changed through dissolution and diffusion. From the view of contacts between atoms, a phase interface is an atomic interface between heterogeneous atoms. If some material properties can be predicted according to the contact between "two atoms" starting from the Periodic Table, it will be of interest for atomic scale engineering in the future.

Therefore I calculated those equilibrium electron densities listed in Appendix B according to TFDC electron density data, first introduced and compiled by Professor Cheng Kaijia. Let all of the solid state elements in the Periodic Table be paired up in twos, the equilibrium electron density is calculated using a empirical lever law according to the cohesive energies. These equilibrium electron densities were applied to revising Vegard's Law for unlimited solid solution in Chapter 1.

These equilibrium electron density data can also be applied to intermetallic compounds, assuming we know which atoms are contacts with each other, i.e. atoms that are nearest neighbors.

Actually, identifying nearest neighbors in intermetallic compounds is also an important question in Empirical Electron Theory of Solid and Molecules. For obtaining data about chemical bond network (including bond length, atoms connected by the bond, and the coordination number of each atom), I have been researching and developing a software package, Atomic Environment Calculation (Mater. Sci. Forum, 689, 245-254(2011)). It not only calculates various data of the bond network, also can optimize atomic coordinates from experiments.

This work was supported by China Petroleum & Chemical Corporation under contract No. X504018.

I am currently working on applying those data in Appendix B on intermetallic compounds with the help of Atomic Environment Calculation.

I wish to acknowledge the influence of Professor Chen Nanping, who had always encouraged me and supervised my PhD dissertation on superplasticity. My thanks are also due to Professor Jin Dashen, Director of Metallic Materials Branch in Department of Engineering and Materials Sciences of NSFC, who introduced me to Professor Cheng Kaijia. Special thanks go to Professor Cheng, member of CAS, who gave me much help and encouragement in my study of the electron theory of alloy.

I extend special thanks to Professor H. Conrad at NCSU, Professors Tao Kun, Gu Jialin, Bai Bingzhe, and Ms. Chen Shunying at Tsinghua University.

Finally I wish to thank my wife and many personal friends who indirectly contributed to this book through their love and friendship during the past years.

Li Shichun

October 2015